Opening to Abundance

Clearing Limitations
to Grow in Joy, Love, and Bliss

Richard H. Lee
Etheric Sludge Scientist

First Edition

ISBN: 978-1-304-33477-0

Printed in the United States of America

DEDICATION

To Alma...From Alma

Table of Contents

III. Personnel Immersion

IV. A New Perspective

V. Descending

Forward

Opening to Abundance is a book that I thoroughly enjoyed, not only because it helped me upon my own Spiritual journey allowing me to step into my own power, but gave me 'scientific proof,' about the power of Qi, 'Vital Life Force Energy,' and the realization and awareness that Etheric Sludge is real and prominent not only upon the Earth, but with people, places, and things. This book opened my eyes to my own self-limiting beliefs. I have become aware that this is why I couldn't seem to step into my power to fulfill my Soul's purpose for incarnation.

When Richard asked me to help him with his book, I was dumbfounded. Who am I to be involved with such an extraordinary book of valuable information? I wasn't sure why he was asking me, however, I jumped right into the opportunity to see a Scientist's point of view of Spirituality. I have been studying and teaching principles of divinity and spirituality for many years so I was curious if Richard had found 'proof' that the limitations, thoughts, and anger I was experiencing in my own life were being caused by something I could not see.

While reading the book and following his journey, I realized that I too was not allowing myself to step fully into my own Divinity. I was met with much synchronicity as I read, edited, and studied each chapter. I was tested mentally, physically, and spiritually, not only by the material, but by my own ego. I practiced the energy enhancing Qigong exercises and the exercises to pull back my power from self-limiting beliefs. I also used the Infratonic S and Scalene Light as described in the book. After just a few days, I felt much lighter and more present. The pains in my sacrum, solar plexus, and heart chakra had disappeared, and I was able to hear my Higher Self much more clearly.

By the end of the book, I had realized that I am the one who creates my world and my reality. I realized that I was my own enemy, getting in my own way on the path to the Abundance, happiness, joy, and bliss, which Jesus Christ stated was a gift for all.

I proudly recommend this book to anyone who is seeking 'proof' that Spirituality is authentic. Many people throughout the world have lost faith in religion and Spirituality because of the lack of knowledge, wisdom, and proof that it is real. Some people, like Richard and me, must 'see to believe' and this book will give you exactly what you are looking for........proof that God is everywhere, inside and all around you. This book will pave the way for you to fully open your heart to Abundance, Joy, Love, and Bliss.

~Blessed Be~

Valentina Marie,
Ordained Minister,
Doctor of Metaphysics,
Doctor of Divinity
Reiki Master

Preface

Welcome to the world of a scientist who made the mistake of studying vitality. Most scientists have the common sense to avoid anything that smacks of vitality, intentionality, or any of those unscientific fields. They believe that they will ruin their careers because no self-respecting peer-reviewed journal will ever publish their work. Unconsciously they fear that they don't know where it will lead, and it might take them to places that are dangerous, uncomfortable, or just different.

My exploration started innocently enough. I was studying bad energy, ill-humor, bad Qi (Chi), which I named Etheric Sludge. It seemed distant and safe. I could study it as a scientist, breaking it into pieces and studying them one by one. When I started writing this book, the title was **Etheric Sludge Science**. As I went further, I discovered that it was important to point you, my beloved Reader, not to the dark walls of the tunnel, but to the Light at the end of the tunnel. I struggled, but the name finally changed as I realized I was awakening from limitation and **Opening to Abundance**.

In Part I, A Scientist Explores Subtle Energies, I am that innocent scientist, safely on the other side of a thick, impervious glass wall. The scientist in you will enjoy the exploration, analysis, and all the unconventional discoveries and theories. However, if you identify with being a scientist, be careful. There are so many "findings" in the first part of this book which, if pursued, will take you way outside the boundaries of science, for you see, science, as it is practiced in the modern industrial world, is built with these same glass walls that separate the observer from the observed. Most of the seemingly objective "findings" you will find in part one are laden with consciousness, that "substance" which is supposed to stay on the other side of the glass wall.

Part II, Etheric Sludge Science, again seems deceptively safe. I am simply studying a machine, an electromechanical device which is purported to produce emitted Qi. The further I went, the more I could feel that I was standing on the outside looking in. At that time, I owned Chi Institute, a research organization that studied Qi. We taught courses to doctors and therapists and took them on trips to China to study "Qigong," which means Qi exercise. These people who joined us were different. They glowed. There was something magical about them. Kirlian photographs of their fingertips showed that almost all of them had what I called the "smooth glow" or the bagel. Here is a photo of my finger compared to theirs.

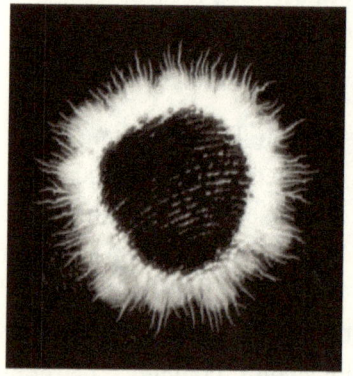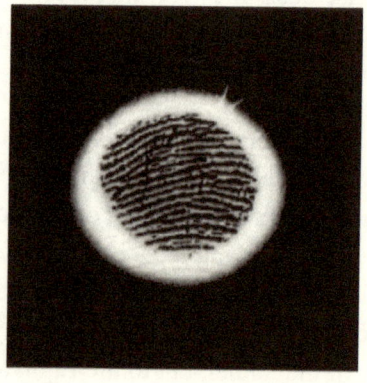

My Kirlian Photograph   Their Kirlian Photographs

There were trips to China during which I was the only person out of 30 or 40 people on the trip who had the balls and streamers instead of the smooth glow. I definitely felt singled out!

My picture shows electrons behaving as they should, looking like lightning as the negative pulse brought electrons into my fingers, and pooling into balls as the positive pulse pushed them away. They were nicely compartmentalized into areas of light and dark, just like my mind. This was also my way of experiencing and studying the world at a safe distance. However, these people who traveled with me displayed "quantum" Kirlian photos. The electrons around their fingers flowed smoothly and illuminated the air as waves instead of particles, smoothly filling the Kirlian picture with a soft continuous light. I discovered that these people were all healers. They lived in a world of smoothness, connectedness, very different from my world of subject/object scientific compartmentalization. My mother played the Cello, and the fingers of her right hand, which held the bow was all balls and streamers like mine. However, her left hand, which held the strings and wiggled and moved so gracefully, adding depth and grace to each note, showed the smooth glow. Every single finger!

I wanted what my tour participants had, and discovered that, through meditation and deep relaxation, my Kirlian photos began to change toward the smooth glow. This is dangerous territory for a scientist. Fortunately, I discovered that I could return to the balls and streamers through caffeine, TV news, conflict, and intellectual discussions. Remarkable!

In **Part III, Personal Immersion**, I found myself inexorably drawn into a strange world of direct experience. It was not like I could simply dive into the water and swim as a fish. I continued to wear the trappings of the scientist, talking and writing in terms of science. But I could feel that the veil between me and my world was crumbling, first into chunks, then pebbles, then sand. My protective glass house became uncomfortable. I realized I had always been uncomfortable behind this wall, and longed to be free of it. I started to build devices like the Scalene Light (Chapter 10) and the Ra-Ad (Chapter 12), which would simply dissolve the walls of the house, so I could be free of this "residual sludge" that I found between me and my world.

I came to see those balls and streamers and the darkness between them as representing, or maybe literally being, the obstructions.

Part IV, A New Perspective, was a HUGE step for me. I realized that the rocks of broken glass between me and my world, my etheric sludge, were my limiting beliefs, traumatic memories, cultural beliefs, and ancestral traditions, which resonated with me and seemed to possess me. I discovered that I could simply "pull my power back from these limiting beliefs," and the rocks which had stood between me and my

11

world would turn to dust. Amazingly, the Scalene Light and pulling back my power, radically different technologies, accomplished similar results.

I was rapidly becoming free of my glass house, and in **Part V, Descending**, I discovered that my purpose was more than being an Etheric Sludge Scientist. I discovered I had been working all these years to help build the bridge from limitation to abundance.

We are all building this bridge together. Each of us is trapped and held by our own demons, traumas, cultural baggage, or dogmatic beliefs. Each of us needs a different flavor of the bridge, to cross to the freedom on the other side.

Opening to Abundance is a bridge for the part of you that believes you need science or other structure in your life to protect you from direct experience of your world. Follow me now on this pathway. When you get to the other side, you will find yourself living in a world that is made of nothing but Joy, Love, and Bliss, waiting for you to dissolve the separation, the limitation that stands between you and abundance.

Enjoy the Ride!

> Richard H. Lee
> Laguna Beach, California
> August 24, 2013, 4:44 AM
> My Brother's Birthday.

Acknowledgments

Acknowledge: To publicly recognize a contribution.
Appreciate: To increase the value of…

The role of a co-creator includes the powerful shamanic practice of appreciation: increasing the value of all that is around us through the quantum act of noticing with heartfelt appreciation.

I truly appreciate the gifts of intelligence, awareness, concern for detail, the love of nature, and knowledge of the importance of diet, which I have learned from my mother and father. My parents recently passed and I have been living in their home for the past year. My growth in confidence and maturity has been immense as I stepped into my father's confident and thorough nature, learning to care for his garden.

Huge appreciation goes out to Valentina Marie. I met her through The Soul Intention Spiritual Fellowship, and received an amazing energy healing session from her. I asked her if she could help me assemble a book from the collection of bits and pieces of writing I had done over the years. She obliged, jumped right in, and was truly inspired by the material. Within weeks, Valentina carved out a 400+ page book. She has brought a great thirst for wisdom, truth, and knowledge on top of what is already a huge degree of Spiritual knowing regarding the metaphysical side of this book.

I couldn't have completed this without my staff at Sound Vitality, Karene, Mina, Maria José, and Dan, and their staff of consultants, who have been spectacular at holding the company together during these past two years of my rebirth and awakening, allowing me the freedom to evolve and transform in unexpected yet beautiful ways.

I appreciate Master Wan, Master Duan, Samuel, Sai Baba, my brother John, Sydney, Dr. Fu, Patricia, Ute, Suzanne, and all my other teachers over the years, who have taught me so much and enriched my life so fully. I appreciate Maria José, Fiona, my brothers Bill, John, and Tom, Catherine, Marie Nelson, Dr. Bota, Ron Riegel, Laila, Jerry Alan, Melinda and Emily, and so many others for embracing huge chunks of my dream by appreciating and loving me. Their appreciation and support have provided a huge foundation for my growth.

I appreciate the huge support from the Heavenly hosts of unseen beings who love me and hold me unwaveringly: Jesus, Mary Magdalene, St. Germain, Archangels Michael and Raphael, and all my unknown and forgotten guides and teachers. I appreciate my Higher Self for creating such an amazing adventure as I continue to walk this path of life before me. And I appreciate Alma for your unwavering support, guidance, and participation in our co-creations.

I appreciate all mentioned above, and all of you whom I didn't mention, who have enriched my life and contributed to this book in so many ways.

I know I am a Spiritual student upon this path, so I get to appreciate all of you for helping me in my process of becoming a remarkable being as I continue to grow two, ten, one hundred years from now. I appreciate you, my Reader, for appreciating whatever it is you are seeking and can apply to your own life. I appreciate you, for it is your appreciation that gives this book life.

Abundant blessings to each and every one of you. May you bring Joy, Love, Bliss, Prosperity, and Abundance to every heart you touch.

I. A Scientist Explores Subtle Energies

1. The Beginning of the Journey

"If it is real, it must be Measurable"

Childhood and Early Life

When I was a little kid, Albert Einstein was my hero. When people asked, "What are you going to do when you grow up?" I answered, "I'm going to change the world," with no idea what that might mean. As a grade school child I could fix most anything around the house. I seemed to have a knack for seeing into devices, and knowing how they worked before I opened them up to fix them. Also, I could find virtually anything that anybody lost in the house. It was a bit embarrassing when I would open my mother's lingerie drawer to find her lost keys.

When I was 13, I was already a bit of a scientist. My brother John, older by 5 years, had taken another path. He had started Transcendental Meditation and had lots of crazy ideas. For instance, he told me there were rivers that flowed deep underground and could be flowing under his bed, so he had aluminum foil under his bed to reflect the energy so his energy wouldn't be drained by the underground rivers. (I later discovered that dowsers can generally accurately find underground rivers by detecting the emanated energy using dowsing rods.) He would also put jars of water out in the sun with purple plastic wrapped around them to charge them with violet energy because he said he needed more violet energy in his body. He also had an infuriating habit of saying, "Everything is energy." He would talk about *prana*, a form of life force energy that was pretty magical in what he claimed it could do, and was the basis of human health and vitality. It seemed to me at the time that the perspective that "everything is energy" was a whitewash to ignore the scientific implications of things and to obscure the fact that scientific investigation rejected these notions of *prana*. In my long, heated conversations with my brother, I often insisted, "If this stuff, this energy you talk about, is real, it must be measurable!"

These discussions with my brother, and particularly the idea that "if it is real, it must be measurable" have pretty much dominated my life over several decades, as I have come across vast subjective evidence of the existence and importance of these unseen and apparently un-measurable energies. I have built or bought every kind of energy sensor I can imagine, and have measured lots of things I'd never heard of before. Still I found electronic sensors and computerized analysis equipment hopelessly inadequate, and definitely inferior to the detection and discernment ability of many highly sensitive people, such as my brother.

After graduating from Harvey Mudd College, one of the Claremont colleges, I found myself the only graduate who neither had a job nor was going to graduate school. While my classmates were earning $40,000-$50,000 per year, I landed a job with the Claremont Colleges for $11,000 a year working in their maintenance plant improving the energy efficiency of the buildings there. Efficient energy use was very important to me. This continues today. With the devices I develop and put into production, I am insistent about making them as energy-efficient as possible. (My Infratonics can run for 60 hours on a single charge, and my Vital 1 can go 400 hours on a single coin cell.) So after I started my energy research and before I assumed the role of Etheric Sludge Scientist, I called myself an Energy Systems Engineer, who just happened to be focusing on human energy. Efficient use of human energy became as important to me as efficient use of electrical energy had been. I soon discovered that the main obstacle to efficient human energy use was Etheric Sludge. Whether called heavy energy, bad Qi, bad karma, DOR (dead orgone), oppressive energy, pernicious influence, Mercury in retrograde, or anything else, anything that interfered with the efficient use of human energy became "Etheric Sludge," the object of my pursuit.

Etheric refers to the substance of the energy field, or aura that surrounds our bodies. Thus, Etheric Sludge is that goop in our energy fields.

Picture of Etheric Sludge

Etheric Sludge, as defined by me, is a broad, scientific term which has come to mean anything that comes between us and our experience of joy, love, and bliss. As we take a closer look we find that Etheric Sludge takes on a wide variety of forms. It is the stuff that gets into our heads to cause headaches, worry, and depression. It gets into our hearts to cause heart disease, stress, and broken hearts, and it gets into our nervous system to cause conflict, pain, chronic illness, and anxiety. We're talking many flavors of Etheric Sludge. All have in common that they are vibration or consciousness in particular substances, whether electrical, magnetic, or gravitational. Sludge refers that gooey black stuff that messes up physical substances and sinks to the bottom of bodies of water.

The study of Etheric Sludge is the study of virtually everything that interferes with Heaven on Earth, but from a perspective that physics can understand and test. As you will see, my pursuit has led me on a long meandering journey, to many lands, philosophies, and experiments. I wanted to call this book "Etheric Sludge Science," however my editor and my higher guidance had a different plan. For, you see, while my search has seemed to be to study Etheric Sludge, in fact, it has been to free ourselves from its limitation. The search for Etheric Sludge was my search for that which limits us from experiencing or inheriting abundance. As it turns out, we can spend an entire lifetime studying Etheric Sludge, forgetting that our real purpose is to remove the limitation of Etheric Sludge so we can open to abundance.

The World of Traditional Chinese Medicine

I have been told repeatedly that there is no way to understand Qi except through direct experience. Like almost any modern scientist, I ignored this advice, convinced that there is only one reality, with scientific, objective, repeatable observation as one window, and direct, subjective experience as the other window. The result of the subjective path is self-knowledge and growing personal potential, a search for personal purpose and meaning. The result of the objective path is effective tools that anyone can use to change the world in powerful, predictable ways. Reality is the same regardless of viewpoint; but because our perspectives are different, our observations are often in conflict.

In my search for understanding, I discovered that many of my fundamental convictions were challenged. You too will probably find matters in this book that may cause you to question your own beliefs. Simply take notes and wait. That which reflects the real world will ring true and be verified by your personal experiences. Much of what I present is not yet proven or unproven. It must be tested by science. If I have my way, this testing will come about in the near future.

My first foray beyond the horizon of my scientific world came when my parents invited me to accompany them on a trip to China in 1982 soon after its doors opened to tourists. While I found the Chinese people quite different, they seem to be equally observant of the blonde hair and imponderably large bellies, bottoms, breasts, and noses of our American tour group. They thought cow's milk causes these strange phenomena. Now that cows are proliferating throughout China and meat and dairy products are in abundant supply to those who can afford them, a rapid fattening up of city dwellers in China may be proving that theory correct.

On our tour, fifteen of us visited the medical clinic of an agricultural commune in Shanghai. It was housed in a two-story concrete building with a flat roof, approached by a packed earth driveway. Each step we took toward the building raised dust. This was not at all my idea of the ideal sterile environment required for medical treatment. We climbed the steps to the second floor, where we packed into a small room to observe.

One man was lying on his stomach as a doctor took a 3-inch-diameter glass cup, waved a flaming ball of cotton dipped in alcohol inside it then pressed the cup to the patient's back. As the air inside the cup cooled, a vacuum was created that sucked the skin into a bubble we could see under the glass. The doctor applied another cup and explained that this process was called "cupping." It drew stagnant blood and Qi from the point where the man felt pain up to the surface of the skin. When the doctor removed the cups, round circles of red and purple discoloration, like colorful bruises, remained on the patient's back. The patient would proudly show these awful-looking marks to his friends as proof that the doctor's therapy was successful!

An older woman was sitting on a chair with three-inch wires protruding from her arm and shoulder. At the end of each wire was a one-inch diameter ball of gray material that the doctor was setting on fire. He explained that the energy from the burning ball would travel down the wire and warm an area deep within her shoulder and arm. "I had 8 treatments last autumn and had no trouble with arthritis all winter," the patient proudly proclaimed. The doctor nodded in agreement. The translator explained that this was acupuncture.

Could this be real, I wondered. Or were these just government employees playing the Communist Party line for foreigners?

The second and only other room in the clinic was even smaller, full of drawers, bottles and jars. The doctor showed beetle skins, dried snakes, and scorpions. He explained that the proper mixture of these ingredients, along with Chinese herbs, could cure almost any disease.

That was it, the entire clinic. We saw nothing that I, as a Westerner, could associate with medicine. No antibiotics, no antiseptics, no surgery, no sterile dressings. We retired to the dining room for a 20-course lunch of fresh, delicious stir-fried vegetables. "How is this possible," I thought, "that those people believe in such crazy things? And how is it possible that they stay healthy without antibiotics and big white gauze bandages?" I concluded that this commune was a backward place that relied on faith healing. But a question lingered: "Was it possible that some aspect of what they were doing had some medical value?" Certainly these patients believed in it. And the Chinese government featured it on our tour. Could a quarter of the world's population really be so foolish as to believe in smoking balls and scorpions?

With my engineering outlook, I reasoned that if it did really work, there must be a scientific basis for it. My world could not, however have been more different from that which I had seen in the Shanghai clinic. Nothing allowed for this alternative reality. I had never read anything that hinted at scientific evidence for the value of these Chinese therapies. Without an experience that would confirm the efficacy of the treatments, the scorpions and smoking balls remained just oddities that I had seen on the other side of the world.

Electricity in my Leg

One bright Saturday morning in 1983 I set out for a run from my house in West Los Angeles. While running toward the ocean along Venice Boulevard, I came across a sign that said Ryokan College. I had seen that sign before, but had never thought anything of it. This time, perhaps because I was tired, I stopped and looked at the building. It probably held three or four rooms. The closest room had a sign in the window, "Acupressure Treatments - by Appointment." The door was open, so I walked in.

The therapist was very kind, taking time to explain to me some of the principles of energetic acupressure. Her system of acupressure, Jin Shin, worked by activating "strange flows" and allowing the energy in the body to flow from where there was too much (causing inflammation and stagnation) to where there was too little (causing weakness and degeneration). This of course sounded strange to me. It also sounded a little too simple. But she was kind and the office was very clean and "medical looking," so I agreed to an appointment.

The patient room was also clean, with pleasant, quiet music playing in the background and a massage table in the middle of the room. I took off my shoes and lay on my back on the table. She put her fingers on points on my neck, head, and shoulders. Sometimes I felt a pulse at those points, sometimes not. I did feel that she really loved what she was doing, for I felt growing love or harmony as she held points on my neck and head.

By the end of the treatment I had felt nothing out of the ordinary, but I was impressed by her caring attitude. She explained that my energy was constrained, not flowing freely. This, she explained, was why I didn't feel the energy flowing in my strange flows. I discounted her ideas of constrained energy and attributed my failure to experience the "strange flows" to my non-suggestible, scientific training and nature. I believed that there was nothing there to feel. But, just in case my interpretation was wrong (and probably to experience more of her loving harmony), I signed up for another treatment the following week.

During the second appointment, the acupressurist explained that my shoulder tension was caused by stagnant energy in my liver, and that we would work on freeing up that energy. For 30 minutes she worked on opening up the flow of energy in my body. Then, while she was holding one point on the left side of my neck and another point on the right side of my body at about heart level, I felt a tickling down my left leg. It grew until it was a strong, distinct ribbon of electrical tingling flowing from my hip down the outside of my leg, through my ankle and foot to my fourth toe. It felt rather like the goose bumps I sometimes feel up my back when I hear something inspiring, except that it was concentrated in one thin ribbon down the outside of my leg.

She had made no comments or suggestions while I was experiencing the strange sensation. After it stopped, she asked, "Has the flow stopped now?" Amazed, I asked how she knew that I had felt anything. She explained that she had felt the current flowing through her fingers, and when it stopped, she knew that a new balance had been established.

Had I imagined the sensation? Did I want so much to experience something that I had created it? And why did I feel the current in my leg while the acupressurist's fingers were on my upper body? How did she know when I was experiencing it? How was she able to experience it with me?

I knew the Chinese definition of this flow. It was Qi I had felt flowing within me. But this was not a satisfactory answer at all. Qi was something theoretical, that I did not really believe in. The flow felt like electricity. But without a battery, how can electricity flow in the body, and why did it flow where it did? If this balancing the acupressurist talked about was really important, what was flowing? What was the nature of this balance? Were electrons moving from my hip to my toe? Or from my toe to my hip? How were they stored? How did acupressure establish this new balance? My acupressure experience did not answer any of my questions. It created 100 more.

Dr. Fu

A year passed before I met the answer. A coworker's wife had invited three doctors from China for Thanksgiving dinner, and since I had been to China once, they thought I might be able to help entertain. I was the last guest to arrive, and the three, all women, introduced themselves. All wore what I have since learned are "Chinese doctor" glasses, with thick black plastic across the top and silver rims around the bottom. Dr. Shi was a short, suspicious-looking woman, a gynecologist,

22

who studied me carefully. I wanted to start a conversation so I asked what she thought of acupuncture. She stiffened, "Why ride a bicycle when you can drive a car?" she pronounced. That didn't go very well. Dr. Jin was a tall, gregarious ophthalmologist, and Dr. Fu was a quiet epidemiologist with a pale face and pink cheeks. None of them would admit to knowing anything about acupuncture. All were proud Western medical doctors, and none engaged in the backward ways.

As it turned out, Dr. Fu was an excellent ping pong player, and we spent hours that night hitting the ball back and forth. I was fascinated with Dr. Fu, because she had such a strange way of looking at things. Whenever I would go off on one of my scientific explanations of things, which I thought were very enlightening, she would simply say, "Doesn't matter." I found that quite frustrating because I thought what I had to say was important. I realize now that most of what I said, and most of what everybody else says, is just words, Etheric Sludge, mostly words repeated from what somebody else said, or what the news said, or nowadays, what the Internet says. So, if you think this book is just a bunch of words, you're right. Consider it a sign that you may be an enlightened being. The whole point of this book is simply that Etheric Sludge is bad and getting rid of it is good, and getting rid of Etheric Sludge is the way you become an enlightened being and experience abundance. There you go. The rest of it is thousands upon thousands of words. However, I think my words are important, so I will continue:

I later had discussions with Dr. Fu about many things, including acupuncture, which, I discovered, she used for her own stiff neck, but didn't tell anyone at USC Medical Center, where she worked. Her patients in China received only the latest Western treatment. As I later walked with Dr. Fu to her apartment we met a neighbor, a heavy Latino woman struggling up the stairs. Dr. Fu explained that this woman had severe low back pain. She helped the woman to her apartment, quickly fetched a small package from her own apartment, and invited me into her neighbor's home. The woman lay down on the couch with her low back exposed. Dr. Fu sterilized her back and produced some eight-inch acupuncture needles. I watched in amazement as she inserted them all the way into the woman's lower back, eliciting no sign of pain or discomfort.

Dr. Fu turned to me and said "Ask her if it hurts." I then realized that these two people, doctor and patient, did not even speak the same language. I asked the question in Spanish and translated the answer, "No pain."

I wondered how this woman ever agreed to allow someone she could not even talk with to stick needles in her back. Did she even know what was happening? I was amazed that the needles were all the way in. If this woman were as thin as Dr. Fu, the needles would stick out the other side.

"Come, feel the Qi," Dr. Fu said to me, twisting the head of the needle. At first, when I touched the needle, it rotated easily, as if inserted into gelatin. Then as I continued to spin it, the needle seemed to be grabbed firmly by the tissue. A very strange feeling. Eight inches inside this woman's lower back, I had felt this needle activate the Qi. It was spooky.

2. Synchronicity and Infrasound

"Life is effortless when we follow the synchronicities."

I went on to marry Dr. Fu and we moved to San Clemente and rented a storefront facility to open an acupuncture clinic. By day we hammered and painted, and by night we slept in a little room in the back with all of our belongings. Synchronicity was new to me, but I learned working side by side with Dr. Fu.

I had never built walls before, but I knew I could buy wood and erect some sort of partition. As I was about to drive to the lumberyard, a neighbor walked up and said "I'm a drywall specialist. Do you need any walls built?" I was amazed that his offer came exactly when it did, but then put it out of my mind. Another day, we discussed having a merchant account so that we could accept payments by credit card. I explained that without a retail background, it would be difficult. Just then someone walked in from a local bank offering to set up a new merchant account for us.

Inevitably, when Dr. Fu was low on energy or not feeling well, patients would call up and cancel – eight or ten out of a schedule of 30 – giving her a much needed respite. When she felt better, the phone would unaccountably ring with patients begging for an appointment. After one busy morning, she looked at her slow afternoon schedule and said to her office manager, "I wish this afternoon were a little busier." By the time she returned from lunch, five patients had called for emergency appointments, and our waiting room was full throughout the afternoon.

Dr. Fu's life was filled with these coincidences, and while I tended to view them as random chance, I had come to expect them. Synchronicity is part of Chinese philosophy and definitely part of being a Qigong Master. Whenever we met a new Master, it always seems the words "Yo yuan" are repeated, which means something like "The heavens have arranged that we should meet today."

Synchronicity is a basic tenet among Qigong Masters. I have found that life is difficult if I try to go against the flow of synchronistic events and much easier if I choose to go with the flow. Yuan is a coherence between the individual and the universe wherein the random chance of the universe is concentrated into occurrences that guide us if we are ready to follow, and make us miserable if we are not! Those who perceive this flow can tell the future. Those who have learned to direct this flow can work magic.

Synchronicity, cooperation with unseen forces, is a major theme throughout this book. Read on and you will discover that life is much more than what it seems.

The Qigong Therapy Apparatus

These synchronistic events continued. One day, a month after the clinic opened, Dr. Fu received her green card, which meant she had become a legal resident of the United States and could travel. Since she had not been back to China for 5 years, she felt that she must visit China immediately and celebrate the Chinese New Year with relatives who would come from across the country to her parents' home. I suggested that we not abandon our month-old business, but she set about closing the clinic for a month.

Just before we left for China I was organizing the magazines and thumbing aimlessly through one of Dr. Fu's Chinese-language acupuncture journals and found just three words in English - Qigong Therapy Apparatus." My mind began to spin. Could this device possibly reproduce the emitted Qi of a Master? Could it somehow measure emitted Qi? Either way, it would be extremely valuable to science. What if this really were Qigong technology based on scientific research? Perhaps by taking it apart, I could discover the secret of the mysterious electric flow that I had felt in my leg. Perhaps this machine held the secret of Qi. Looking back, I realize that this random discovery was a serendipitous, life-changing moment in my life. It just so happened that we would be passing through the exact city in southern China where this device was manufactured. I began to share Dr. Fu's enthusiasm for the trip.

I waited impatiently to tell her about this wonderful device. Did she think it could produce Qi? But she had no time for such nonsense; she was a Doctor of traditional Chinese medicine, and knew that emitted Qi was human energy. Not only was the device probably fraudulent, but it was an offense to healers to suggest that their healing ability could be reproduced by a machine. To her mind, I just wanted a new toy. She was right, and I was determined to have my "toy" - or she would not have a new husband to show off to relatives in China!

On our way to her home in China, we stopped by the factory to pick up a "toy" for me. It was an odd thing. It was made of faded yellow plastic which made it look 30 years old. I plugged it into the 220-volt outlet in our Chinese hotel room, and it wiggled with an irregular pattern. Was it just a glorified massager? The box said it had been used successfully in hospitals to treat tracheitis, bronchitis, and uterine bleeding, among other diseases, as well as in surgical recovery. It claimed that it reproduced the emitted Qi of Qigong Masters.

According to the instruction manual, "Medical Infrasonics" was the basis of this device. The inventor's original research, performed at China's National Institute of TV and Electro Acoustics, had shown that very low-frequency sound is an important part of the emitted Qi of Qigong Masters, and this infrasonic sound (so low that you cannot hear it) was effective at increasing the energy level of hospital patients and accelerating their recovery. Her findings were verified by over 100 laboratory experiments and with many of the most powerful Qigong Masters in China. More than 1,000 patients in Chinese hospitals participated in studies, which were claimed to show effectiveness for the following conditions: "Peri-arthritis of shoulder, sprain, cervical vertebra, dysmenorrhea, pharyngitis and laryngitis, gastric ulcer, coronary heart disease, external hemorrhoid, functional disorder of the intestinal tract after abdominal surgery, arthritis, sciatica, women's hemorrhage from use of IUD, pelvic inflammation, pediatric tracheitis, chronic gastritis, sudden deafness, and high blood pressure. The claims seemed absolutely absurd. How could a "massager" perform such miracles?

Personal Verification

I tried it on myself and did not feel anything but an irregular massage action. Since nobody around me had bronchitis, broken bones or surgical complications, there was nothing for me to fix. I decided to take the apparatus apart to look inside for its secret. I spent a few hours searching every store that looked like a hardware store before I finally found a Phillips head screwdriver, then hurried back to the hotel and started removing screws.

The apparatus was full of electronics, lots of integrated circuits. Could the secret of Qigong be hidden in an electronic circuit? I couldn't believe that. There was a transducer, shaped somewhat like a telephone handset, which I unscrewed and found what looked like a speaker assembly made of unlikely components. The cone, usually formed of stiff paper, was made of solid brass, and the diaphragm, usually fashioned from flexible, plastic-coated paper or cloth, was made of rubber. I found a big magnet assembly with a magnetic air-gap to accommodate the brass armature wound with copper wire. Was it an infrasonic speaker? Was it magic? Or was it fraud? I picked up my Phillips head screwdriver and screwed the transducer back together. I had seen no magic. I'd seen no Qi. How could I determine how the crazy thing worked, or if it even worked at all?

I packed up the Qigong apparatus and we set out for Jiangmu, Dr. Fu's hometown. Once again, the wheels of synchronicity solved my problem: We arrived in early February. It had rained for 40 days, and in that time the temperature had not risen more than 8 degrees above freezing. Inside the family house, it was dry and absolutely packed with relatives, but there was no heat. The doors and windows were open all the time to avoid trapping evil spirits in the house. The cold penetrated to my bones and provided me with wonderful cold symptoms and the perfect opportunity to test the apparatus.

First, my headache. My whole head ached. I applied the apparatus and it did help. I could make parts of it disappear with 10 minutes of application. Next, I got an intense sore throat. Within 10 minutes of applying the machine to one side of my neck, the pain on that side vanished, although the other side of my throat still hurt. This was pretty persuasive. I moved the device to the other side, and in another 10 minutes my sore throat had disappeared entirely! This was astounding, but could it be the placebo effect? I tried the machine on my runny nose, but my nose ran even more. Hmm…that could be useful in cases of clogged sinuses.

At that point I didn't know whether to feel misery or ecstasy. I felt that the apparatus was really doing something, so I insisted that others try. Dr. Fu was embarrassed at my insistence that everyone try the apparatus, but went along. Ten minutes later the device had relieved her sister's back pain. I began to believe that there really was some magic in the Qigong apparatus after all. I became even more determined to find it.

"If it is real it is measurable." This appeared to be real, and because it was a simple electromechanical device, it was measurable.

I persuaded Dr. Fu that we should travel to Beijing to meet the inventor at her Acoustics Research Institute. I hoped to discover how the apparatus really worked.

Engineer Lu Measures Emitted Qi

Measuring the infrasonic output from the palm of a Qigong Master is difficult because the world around us is absolutely bursting with infrasonic sound. The experiment must take place in a room that is insulated from all external sounds. Because infrasonic sound travels so easily through all known substances, this is no easy task.

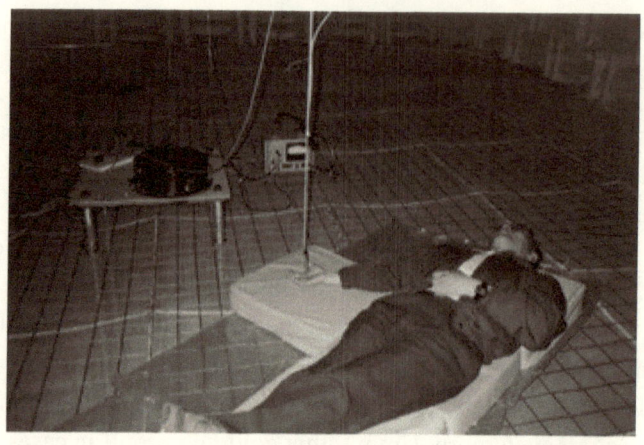

Engineer Lu invited us into a special room through a heavy, sound-absorbing door. The room, about 30 feet square and 20 feet high, was lined with metal to isolate it from any electrical or magnetic influences. The walls, ceiling, and floor were covered with a two-foot- thick layer of sound-absorbing fiberglass "fingers" that reached toward the center of the room to absorb any sound. Wire screening stretched from wall to wall 10 feet above the floor. In the exact center of the room we saw the Qigong Master lying on the wire screen, his hand outstretched, palm up, with a small infrasonic microphone suspended two inches above his palm.

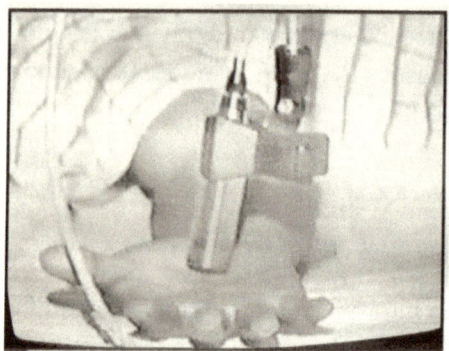

We left the room, leaving the Master alone in silence. The heavy doors were closed behind us and we watched the spectral analysis equipment analyze the infrasonic aspect of his emitted Qi.

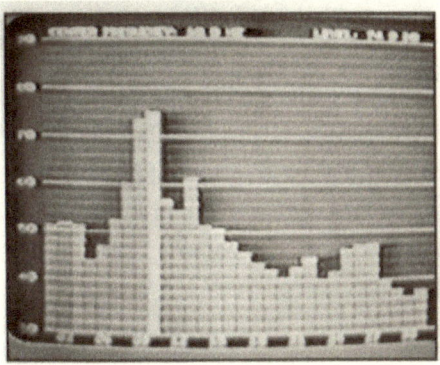

Measuring My Emitted Qi

Everyone is constantly absorbing and emitting infrasonic sound. Studies conducted in this room revealed that the average healthy individual emits a signal strength of about 50dB, and ill or elderly people emitted at a significantly lower level. Most Qigong Masters emitted at a level of 70dB or above. This is 100 times the 50dB signal of the average individual. As I watched, the test equipment showed that this Master had emitted at a strength of 74.5dB, which indicated he would be very effective in treating patients with emitted Qi.

Engineer Lu offered me a chance to see how high a level of emitted Qi I could produce. I rolled up my sleeves and entered the anechoic chamber. As I walked toward the middle of the room I could barely hear the heavy door close behind me. The sound of my footsteps was immediately absorbed by the walls. There were no echoes. I spoke and could barely hear my own voice. I lay down with my hand a few inches below the infrasonic microphone. As it turned out, my emitted Qi was no higher than that of the average individual.

The Infratonic is Born

Engineer Lu explained that she started her research when a friend of her family, a Qigong Master, approached her with his belief that sound was coming from his hand, which he claimed she could measure. He proposed that the sound, when produced by other means, would have an effect similar to the emitted Qi of a Qigong Master. She did not then believe in Qigong and tried to ignore him, but family obligations forced her to honor him by making the first tests. She was amazed at the results, and proceeded to build the original Qigong Information Instruments for testing. When she tested them, she got impressive results. She gave me one of these original prototypes.

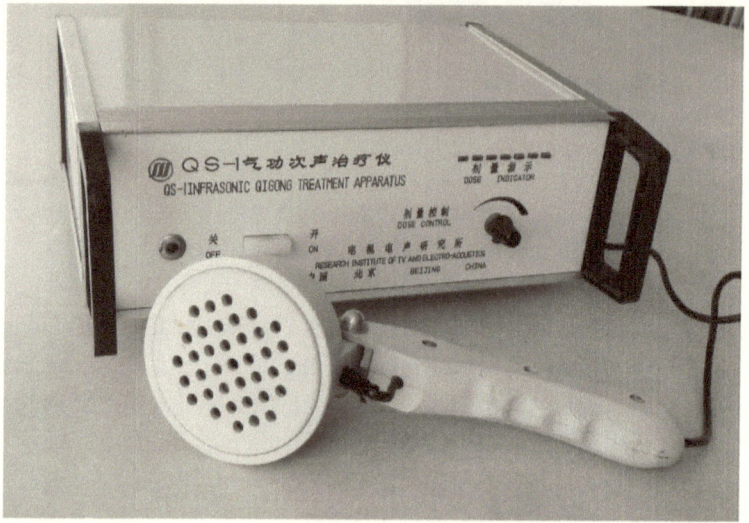

And when she presented these results, the China government financed further research. At that time Deng Xiaoping had recently become chairman of China and sought to determine whether there was any scientifically demonstrable value in the practice of Qigong. He started a program of studying these Masters in high-tech scientific laboratories throughout China. Engineer Lu's project was part of this national effort.

One of the studies she presented to us involved 12 rabbits, all of whom were injected with bacteria that caused a respiratory infection. On the 2nd, 3rd, and 4th days, 6 rabbits in the experimental group were treated with her experimental Infratonic device for 15 minutes twice a day, on both the sternum and the back of the neck. On the following days they were treated just once per day. By the 3rd day, the blood serum antibody level of the group treated by the emitted Qi was 5 times that of the control group. This meant that the immune systems of the rabbits treated by the apparatus were producing antibodies to fight the bacteria 5 times faster than the group that did not receive the Infratonic treatments. By the 7th day their antibody level was 8 times as high as in those who did not receive the treatment. After 21 days, the blood serum antibody count of the control group was still below where the experimental group's count had been after just the first day of treatment, as shown in the graph.

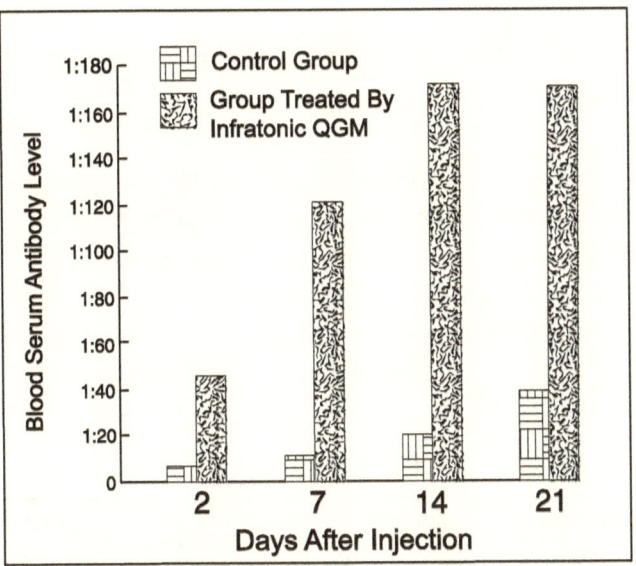

The researchers also monitored fluctuations in body temperature as the rabbits fought off the infection, and found that those in the control group had major fluctuations in temperature, just as humans with infection go through periods of chills and fevers. The rabbits treated with the Infratonic showed a much more consistent body temperature, which indicated a stronger immune response.

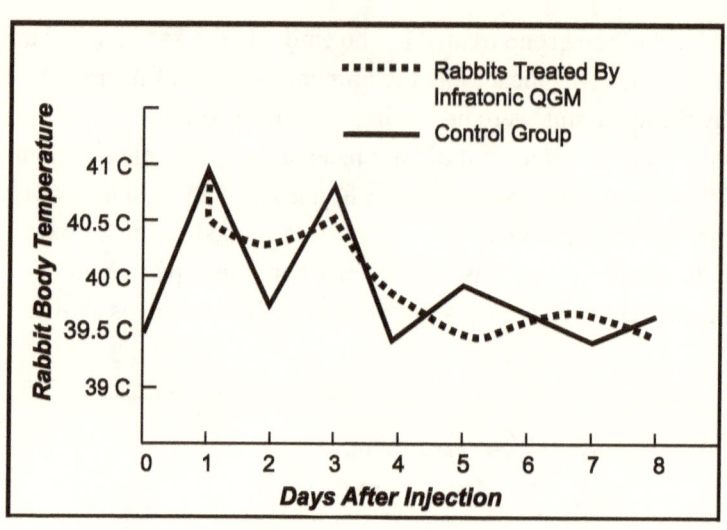

Another study she described, conducted by He Chongnan at the Beijing Institute of Traditional Chinese Medicine, evaluated the effectiveness of the Infratonic device on children with chronic digestive problems. Of the 300 children included in the study, 121 had sustained symptoms for between two weeks and two years, and the other 179 had sustained symptoms for at least two years. The following table summarizes the results, which indicate that about 90% of the symptoms were relieved by the end of the 6th day.

Digestive Problems Treated with Qigong Simulator	Children with listed symptom at start of the study.	Children with listed symptom after 6 days of treatment	% Reduction in Symptoms
Loss of Appetite	300	15	95%
Bloated Stomach	89	8	91%
Aching Stomach	125	10	92%
Diarrhea	25	4	84%
Constipation	150	32	79%
Excessive Thirst	150	39	74%
Pale/Yellow Face	181	115	36%
Distemper	153	24	84%
Insomnia	216	51	76%
Automatic Sweating	61	9	85%
Night Sweats	225	25	89%

Breaking into College in China

There was little chance that Li Wen Chao could ever attend college. As a high school senior, he had achieved only moderate grades in his courses and had done poorly on previous national examinations. Without very high scores in that China College Entrance Examination, Li would almost certainly spend his life in a menial factory or farming job.

The College Entrance Examination was terrifying, because everyone who faces it knows that only half of high school graduates will enter the nation's college system. *Failure in this examination was a common cause of suicide.*

Enter Su Cheng Wu, a medical researcher and physician, and Li's uncle. He had just completed a successful study involving treatment of pediatric bronchitis with the Qigong Simulator and was searching for other things to study. He felt that stress, nervousness, and mental overload were the principal reasons that high school seniors failed the examination, and believed this device would induce the deep relaxation and mental clarity of alpha into the students' minds and bodies, helping them to relax and overcome the mental overload that would otherwise cause their brains to lock up during the test.

For 3 days before the examination, Dr. Su went to visit his nephew's class and treated each student in the class by holding the device on the upper back for 5 minutes and encouraging the student to talk about fears and tensions regarding the upcoming test. For students suffering from symptoms such as headaches, dizziness, poor appetite, insomnia, and menstrual pain, he instead selected and treated a different point, usually along the front midline of the body.

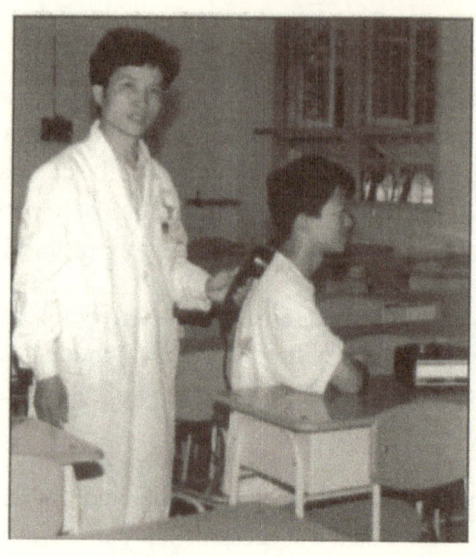

The results were incredible. Li's class was average among 7 senior classes, yet performed the best by far. Where the average pass rate for the other 6 senior classes was 50%, 86% of the students in this class passed the examination and were placed in 5 year colleges! The other 14% got high enough scores that they were accepted to three-year trade schools. Thus, 100% made it to some sort of college.

While it sounds unreasonable that a little relaxation can have such a large increase in scores, the opposing view makes more sense. *Stress and worry can shut down the thinking process of many students.*

Dr. Su became an overnight folk hero, being flooded with gifts from relieved students and ecstatic parents, and leaving administrators scratching their heads. Li Wen Chao enrolled as a freshman at Guangxi Medical University to become a physician.

Engineer Lu had indeed succeeded in producing an effective medical device. In the process, she may have developed a mechanism that emits an aspect of emitted Qi of Qigong Masters that accelerates healing. This is some of the earliest scientific Qigong research, and encouraged funding for other research across the country, including research relating to Qigong and EEG that I will describe later. It also helped support the mounting scientific interest in Qigong research, and led to the First World Conference on Medical Qigong in Beijing in 1988. I didn't find out about this conference until the Second World Conference in 1993, to which I led a delegation of about 70 American doctors and therapists. This American delegation turned out to be about 1/3 of all the attendees, and the vast majority of attendees at Master Wan's Qigong training which followed. In addition, several attendees of this trip got together afterwards and formed the American Qigong Association to forward the practice of Qigong in the USA.

Thus, I had found my passion, a wide-open field of science in which the elusive energies that my brother spoke about so many years before were both real and measurable.

3. Learning to Emit Qi

"Scientists can see and experience a phenomenon, but until they can explain it in terms of their scientific understanding, it has no place in their accepted reality."

The Qigong Conference

The 2nd World Conference for Academic Exchange of Medical Qigong was a 5-day affair with hundreds of papers presented covering scientifically measurable laboratory and clinical research that show the benefits of Qigong exercise, emitted Qi and other phenomena associated with Qigong. This was a multilingual conference, with Chinese, Japanese and English. While understanding the research being presented was often challenging, the abstracts were also published in English, which made the process of studying what was presented much easier.

Western scientists are welcome to close their eyes to facts. The fact is that the research conducted in the major scientific institutes across China and presented at the First and Second World Conference shows us clearly that Qi, vital energy, is a very real and important factor in health. The following are summaries of just a few of the excellent papers, selected because they also show us how emitted Qi and practicing Qigong exercises can help overcome chronic disease around the world.

1.   Qigong exercises improve the results of chemotherapy in cancer. The group practicing Qigong while undergoing chemotherapy showed an improvement in overall health and few of the problems usually associated with this therapy. In contrast, the group that received chemotherapy without practicing Qigong experienced nausea, vomiting, poor appetite, decline in blood cell count, etc.

Read this as you wish. People undergoing chemotherapy need Qi. They also have plenty of time to practice Qigong. Further, those who continue to practice Qigong often have remission from cancer. Cancer can be viewed as a tragic death sentence. It can also be viewed as a grand opportunity to expand human potential, and an opportunity to take our lives into our own hands.

2.   *Qigong exercises and blood pressure.* After one year of Qigong practice, average systolic blood pressure dropped from 156 to 125 and diastolic pressure dropped from 87 to 70 in a group of 37 middle-aged people. In measuring brain health it was found that isoelectric statistical mapping analysis of EEG decreased from 60% to 26%. Qigong training not only stabilizes blood pressure but also retards the degeneration of the heart, brain, and kidneys.

Hypertension, one of the biggest targets of Western medicine, fades away with Qigong exercises while the function of many internal organs normalizes and strengthens. On the other hand, blood pressure pills, the products of a billion-dollar industry, temporarily control blood pressure while damaging organs of the body and creating a wide variety of side effects. Many hypertension patients are ready to embrace the self-improvement exercises of Qigong and the quality of life that comes with them as a health-enhancing alternative to those high-tech little pills.

3.   *The effect of Qigong exercises on the digestive tract.* The number of pathogenic bacteria such as intestinal bacilli, cocci, etc. was lower in the long term in Qigong practitioners than in the control group, while the number of beneficial anaerobes, for instance lactobacilli bifidobacteria and so on, was greater in the Qigong practitioners.

Qigong practice normalizes the intestines. And it goes a lot further than just normalizing intestinal flora. Other studies show that it reduces ulcers, chronic diarrhea, and constipation. Restoring vital energy to the intestines causes them to function normally. Are you aware of how many different kinds of digestive problems are treated with drugs in our society? Does anybody know how many side effects are caused by long-term use of these drugs? Regular Qigong exercises will revolutionize healthcare in America.

4.   *Qigong exercises decreased mortality.* 242 hypertension patients were treated, half with hypertensive drugs, and half with Qigong and hypertensive drugs. 30 years later, mortality in the Qigong group was 25%, whereas that in the control group was 48%.

People who die in the hospital are often so drained of Qi that they can't walk and can hardly speak. They are also often so pumped up on drugs that they are unaware of what's going on around them. Worst of all, they spend this time alone, surrounded only by strangers who expect them to die soon anyway.

Qigong practitioners fill their bodies with vitality, avoiding many of the chronic diseases associated with aging. These practitioners are often fully aware of when death is near, decide when to die, and accept death gracefully when it comes. Perhaps most importantly, they don't spend their life-savings on hospital bills.

5.   *Emitted Qi is effective in treating spinal cord injuries in pigs.* All of the pigs receiving emitted Qi after an artificially induced spinal cord injury could walk after 89 days. None of the control group could stand after the same 89 days.

I watched a videotape of this experiment. None of the pigs could walk just after the induced injury. As the months progressed, the experimental group received the emitted Qi and gradually regained use of their back legs. Some of the pigs still had quite a limp, but even the worst ones in the experimental group looked much better than those in the control group, which made no noticeable progress throughout the period.

This experiment was performed by the Beijing Army Qigong Hospital which specializes in treating spinal cord injuries in people. Wan Sujian, M.D., the teacher of our Qigong training, conducted this research. Many of his students who regularly treat paraplegic and quadriplegic patients used the same techniques to induce flow of Qi within us during the Qigong training I'm about to describe.

Learning to Emit Qi

Confusion was everywhere at the end of the Second World Conference for Academic Exchange on Medical Qigong. Participants were rushing through breakfast only to wait for the bus that would take us from the conference center in downtown Beijing to the Fragrant Hill Hotel in the pine-covered hills an hour to the northwest. When the bus finally arrived, it was not big enough for 50 doctors and all our luggage. One of our guides told us that a second bus would carry the baggage.

As our group boarded the bus, I glanced at the luggage standing unguarded in a big pile in front of the hotel. Someone had to see that the luggage was protected, and our tour guides clearly were not going to, so Dr. Fu and I waved goodbye to the group and went back to sit with the luggage.

Three hours later, we finally arrived at the hotel with the luggage and found our group in confusion. Everyone had been assigned a room, but one room had a big spot on the carpet and another had a view of a brick wall instead of a spectacular view of the tree-covered hills. A third room had a toilet that ran continuously, and a fourth room had a shower that was covered with black mold. Our doctors were trying to resolve these problems with hotel employees, who spoke only halting English and required high-level approvals for any changes. The biggest problem of all was that members of our group could not get their luggage. They swarmed around it, but the hotel employees would not let them claim it. An interpreter explained that the luggage was to be delivered to the rooms and it would be disrespectful to allow guests to carry their own luggage.

I could not see that the delivery was in the offing, and the Qigong workshop was about to start. Like so many problems for foreigners in China, the solution, while obscure, was simple. I adopted a serious look, walked over to the chief suitcase guard, and said in a deep voice, "It's okay for these people to carry their own luggage to their rooms." But he did not understand a word of English. Luckily, the interpreter was nearby and translated. There was a flurry of discussion, and the guard stepped back. Our doctors quickly carried their suitcases to their rooms.

Dr. Fu and I hurriedly checked into our room, just in time for the Opening Ceremony of the Qigong Training, a 10 day event that we hoped would be the highlight of our trip. As we walked through the double doors into the training room, I felt an extraordinary shift from rush and chaos to calm clarity. One wall was lined with colorfully dressed Chinese teenagers. Another displayed a huge flag with a yin-yang symbol, with 14-year-olds in Army uniforms standing at attention on either side of it.

The Opening Ceremony

I watched the other participants enter the room, their looks of annoyance and confusion melting into amazed smiles. Thought was suspended. They all watched, enjoyed, and absorbed. We could feel that something wonderful was happening. But what was it? I now realize that it was coherence, or *"Pei He"* established by the powerful Qi and focused intention of these young Qigong students.

The training began with demonstrations by Master Wan's young assistants of several of the Qigong exercises we would learn. Their brightly-colored outfits and the lively Chinese music made a wonderful show. At one point, three of the young men knelt in the middle of the floor, and after deep concentration, simultaneously broke beer bottles over their heads. Then three young women performed similar preparation exercises and with a blow from their hands broke bricks in half. I later inspected those bricks and determined that they were very solid bricks. Even though the young women weighed no more than 100 pounds, with quite delicate-looking hands, I would not want to have any of them angry with me in a dark alley!

Electricity through Our Meridians

Our next step was an initiation and opening of our meridians, the pathways along which the Qi must flow. Instead of using the ancient Chinese techniques, our young hosts chose a high-tech solution. They brought out a long extension cord and plugged it into the wall. Chinese walls have 220V electricity. Each of two students held an electrode in one hand. They then asked for volunteers. I went up, sat on a stool, and they ran electricity through me. One student put a finger on my wrist while another touched a point at the side of my neck. As the 220V power pulsed through me, a sudden muscle contraction jerked my head to the left, and with another touch, another jolt jerked it to the right. The students were regulating this high-voltage flow with their own bodies. It flowed in one arm, through their hearts, and out the other arm on its way to me.

I have puzzled about the dangers of 50Hz of electricity flowing through the hearts of these young students. A friend of mine had built "fibrillators," devices that heart surgeons used to stop the heart in order to test implanted defibrillators. The fibrillator runs a 60Hz current at 5 volts through the heart. This throws the heart into such confusion that it forgets how to beat. It just "fibrillates," trembling randomly until the implanted defibrillator senses the lack of normal pulse and initiates a series of electrical pulses that bring the heart muscles back into coherence. Why did these students not suffer cardiac arrest?

I have read about a Master who can regulate the flow of electricity through a light bulb, apparently varying its brightness by varying the resistance in his body. It seems likely that, by directing more or less Qi into the arms, much as the students directed Qi into their hands to break bricks, a Master can vary the electrical resistance of his body, regulating the flow of electricity. You will see in the next chapter that Qi is electrically conductive.

The students finished with me by touching my "third eye" in the middle of my forehead. As it struck, I saw a blinding light. Perhaps, at that moment, my 3rd eye was opened. Once I recovered from the initial light show, I didn't feel any different.

It was interesting to note that each participant received a different treatment. Some received several touches on the neck, others none, and some received third eye treatment, and occasionally they would receive currents at other points. Apparently the students were able to see or sense blockages in our Qi and were applying the electricity required to balance and open our meridians.

Receiving Emitted Qi

When we had all received the proper current, it was time to open all of our meridians through emitted Qi. This was a remarkable demonstration. Six of us at a time would sit on stools in the middle of the room and two students would treat us individually. One stood in front and one stood behind. We were asked to relax and close our eyes. Then the students would start the treatment by waving their hands around us.

As I watched, some participants started to lean forward or backward. Their eyes were closed. I noticed that the students' movements definitely led the movements of the subjects, and often the students movements were performed behind the backs of the subjects. Soon the room was full of activity, with the students in constant motion with their hands pushing and pulling the invisible Qi. The participants on stools looked like stringed puppets in slow motion. These participants were reputable doctors I had brought with me from the United States, so there was no question they might be acting. Perhaps, of course, because most of the participants saw others do it first, there could have been an aspect of suggestibility in their actions. But still, practitioners located behind their backs were directing their movements.

Some felt violent trembling, others felt heat, tingling, or odd forces. Most felt trust and love. A few felt nothing, and others, like me, could not explain what they had felt. Some had rather violent reactions hours after this initiation. Several had lingering pain in their backs or necks. Others had such symptoms of cleansing as smelly diarrhea or sticky substances in the throat. One older woman had anxiety and chest pain and thought she was having a heart attack. Fortunately everyone's symptoms faded away over the next 2 days. Participants with the most violent reactions seemed to benefit the most from the experience. The woman with the heart symptoms went on to teach Qigong and become an active Qigong healer herself.

As I sat on the stool I felt great trust in the young students even though I had never met them and could barely communicate with them. I closed my eyes and waited. I felt nothing. I heard great activity in the room as a dozen students started their "dance." I waited for something to happen. My brain raced. I felt my own muscles responding. It was not simply the power of suggestion; the only possible explanation, from a Western perspective, was that some unseen signal was sent by the students and received in an unknown way by the brain, which then caused our muscle movement. These students, however, routinely cause paralyzed limbs of quadriplegic patients to move using this process. Healthy nerves between the brain and the muscles are therefore not required for this phenomenon to function. To me, it seemed as if the students were bypassing the brain and directly influencing the nervous system or the muscles.

To a Qigong Master the process is much simpler. Qi is movement in a fluid much like the eddies and flows in a pool of water. Each of us has a Qi body that might be similar to a vortex, an area of swirling water in a pool. The mind usually directs this Qi body, which, in turn, directs the physical body. A Qigong Master can feel these vortices as well as other currents in the Qi, and by waving a hand, can alter the flow of the currents, reshaping the vortex that is that Qi body.

Notice the fascinating lights and colors in this photograph and the previous one. (The e-book has color photos or you can go to www.infratonic.com/abundance) They are exactly as taken by one of those disposable box cameras with no touchups or enhancements. Many other photos showing these sorts of lights were taken by other participants. This seems to be a real, measurable, or at least, photographable effect. From the Qi point of view we are simply currents in the universal Qi body, each in constant interaction with all people and things around us. One of the students tried to explain it to me this way: "Before you have felt Qi there is no way for you to understand it. Once you have experienced it, you know there is no other way to explain the Universe." This is different from the perspective of scientists for whom things are not real until they understand them. They can see them and experience them, but until they can explain a phenomenon in terms of their scientific understanding, it has no place in their accepted reality.

Master Wan is a scientist. He is also a Qigong Master, a corporal in the Army, and a physician trained in Western medicine. He spoke quickly, with enthusiasm, and moved gracefully as he lectured. As a child growing up during the Cultural Revolution, he rescued one of his father's books from the Red Guard, an army of students determined to cleanse China of bourgeois and backward thinking. Young Wan knew that if the Red Guard found the book, they would take it and burn it, then probably punish his family for having it. It was a book on Qigong. With the Red Guard at the door, he escaped out the back with the book, and in the hills among a group of trees, he found a secure hiding place for it. Every day, he would sneak out and study. He never told anyone he was studying Qigong.

He later joined the Army, but kept his Qigong a secret. Then, one day, his superior officer severely injured his leg. Soldier Wan could sense what was wrong and felt his new Qigong skills could help. However, he believed that he risked severe punishment or ridicule if his Qigong secret were made public. He offered to help provided that the officer promise to keep his Qigong skill a secret. It was agreed, and Wan began his emission of Qi and the removal of toxic Qi. Pain relief was immediate and recovery followed. It was difficult to explain such an abrupt recovery, and the news of Master Wan's Qigong ability leaked out. To his surprise, instead of ridicule, he was sent to medical school for a full physician's education. Synchronicity had intervened, and brought him to stand before us at the training.

In 1978, a major earthquake struck Dansang, a few hours away from Beijing, killing half a million people. Army personnel and particularly medical doctors throughout the country were sent to Dansang to help in the recovery process. Master Wan was among this recovery crew. At one time he and two other rescue workers were in a building searching for survivors when an aftershock struck, trapping the three of them in the rubble. Master Wan tried to explain to the other two the principles of Qigong breathing and relaxation and how these techniques could calm and relax them and greatly reduce their oxygen requirement. They tried to follow his instructions but could not overcome their panic. After several hours, both died. Only Wan was alive when help finally arrived. [Panic and anxiety are an important form of Etheric Sludge as you will later see.]

What Master Wan saw in Dansang was suffering. About 50,000 people were paralyzed or partially paralyzed from spinal injuries. If it had been only a few, he with his Qigong could have helped. However he felt helpless facing such an immense disaster. At that moment, his dream of a Qigong hospital devoted to the treatment of paralysis was born.

At first his requests to the Army were denied. But when he was assigned a single assistant in a small room, his dream began to become reality. It was soon apparent that more Qi than he could produce was necessary. He began to recruit young students and to teach them his Qigong skills. He traveled widely, searching for potential students. Many were young teens or preteens who already had good Qigong skills and wanted an opportunity to develop their skills and become more knowledgeable. Their parents turned them over to Master Wan; confident they would be well cared for and would contribute greatly to society. Master Wan's focus on coherence and team-oriented treatment was attractive because many Qigong doctors practiced in isolation and did not fully understand the team approach.

Virgin Qi

Master Wan explained that his treatment program was highly effective because, to a large extent, the students contributed much younger, Virgin Qi. He had everyone's attention! What does 'virginity' have to do with Qigong?

He explained that there was a long tradition of starting students in Qigong before puberty. This was quite important if students were to reach their full potential. Many of his students had been with him for many years, some from the time they were eight years old. Many had started their training at another center, perhaps a temple. Others were trained by their parents who were themselves Qigong doctors. One important aspect of this early training is that young students are able to absorb new concepts, whereas most older students tend to fight them.

My son, Tomu, was fluent in three languages by the time he was three, and at five, he translated among his grandmother (Cantonese), great aunt (Mandarin), and me (English). Children easily grasp new concepts and can often learn in days what it takes adults months or years to master. Unfortunately they often lack the second key requirement in the selection process, discipline.

The ability to learn is not the main reason to start training before puberty. Jing, the inherited Qi that acts as a catalyst for the collection of Qi from the lungs and the digestive system during Qigong exercises, is strongest at onset of puberty. With proper training during this critical time, students can feel the Qi strongly, and Jing can remain strong throughout life. In addition, skills can be learned to build Jing to make it even stronger.

The other factor in starting training early is sexual activity. The onset of sexual activity before the physical body has matured can damage Jing, decreasing the potential development of the individual. Excessive sexual activity at any time in life will decrease Jing, crippling the body's ability to collect Qi and creating such physical liabilities as low back, knee, and ankle weakness. Even a preoccupation with sexual thoughts or sexual desires can deplete Jing because the mind directs the Qi body. A young Qigong student must therefore learn to control his thoughts and feelings regarding sexual activity. "Sealing the Qi," a part of Qigong training at puberty, involves overcoming sexual obsessions through exercises of emotional and mental focus. It involves learning to draw sexual energy into other parts of the body. This both decreases the internal sexual pressure experienced by most teenagers and energizes higher centers for maximum realization of Qigong ability. This is undoubtedly one of the most important concepts in unlocking the power of Qi.

Emitting Qi

At one session, Master Wan taught us how to emit Qi. The concepts are simple, but require a degree of faith in the concept of Qi. Following Master Wan's instructions, I started by relaxing and allowing awareness of the flow of Qi in my body to increase. "If you cannot feel Qi, you cannot control it." Master Wan said, "You must become so sensitive that you can feel the Qi." This requires extensive, quiet, solitary practice. "When you feel coherence with the Qi as your hands move through it," he continued, "then you can begin to work systematically to remove and purify the bad Qi." To feel Qi I would need to become so calm and sensitive that I could feel the flow of Qi within me. To Dr. Fu this was easy, but, although I tried, I found it impossible.

The second step was to control breathing. Master Wan explained that the breath moves Qi. I took a deep, gradual breath and felt the Qi flowing into me through the crown of my head and the base of my tailbone. I drew it into the Dan Tian, the body's main storehouse of Qi, located an inch below the navel. I had been practicing this breathing exercise and thus had an idea of how flowing Qi should feel. I found this earlier preparation very helpful.

The second half of breathing is exhalation. Master Wan instructed, "As you feel the air rising up and out your mouth, feel the Qi also rising from the Dan Tian. Bring it straight up from the lower abdomen through the heart. Smile as it flows upward through the heart and feel it flow across the chest and down the arm through the center of the palm and out. You must direct it to penetrate deeply into the patient, to the area in need of healing."

Master Wan explained that a Qigong doctor must take certain precautions. "First, avoid emitting Qi with both hands." He said. "Keep one palm facing upward to help collect Qi. The important thing here is that you remember to collect Qi during inhalation. There are many Masters who draw Qi through the crown of the head or other points, using both hands to emit Qi. This is risky for beginners. If you stop collecting Qi during inhalation, you will soon have no Qi left. This will leave you both ineffective and exhausted. So, during inhalation, forget the patient and focus on your relationship with the Universe around you. On exhalation turn your attention again to the patient."

Coherence is very important here. Love and compassion strengthen coherence and make the therapy more effective. Master Wan emphasized at length the importance of the attitude of the healer. Cultivating compassion is an important part of becoming an effective Qigong doctor.

Pei He is the term he used. The Master and the patient must be in close attunement, open to the exchange of Qi and working toward the same intention and purpose. Establishing strong *Pei He* is the principal task of mastering any principle, part of why it takes so many years of Qigong practice to become a skilled Qigong Master.

Master Wan cautions Qigong practitioners to be careful not to allow the patient's bad Qi to collect in the practitioner's body. This means first performing extensive Qigong exercises to build up our own Qi. Strong Qi defends against the invasion of bad Qi. The second principle is to keep joints and muscles loose. Never leave a muscle tense or a joint locked because this blocks the flow of Qi and allows bad Qi to accumulate. Continually shift weight from leg to leg, never standing with both feet flat on the floor. Keep the elbows and knees bent. While some Masters treat patients while sitting in a chair, Master Wan felt strongly that this was unwise because, as soon as any part of the body stops moving, in this case the back or pelvis, Qi can stagnate.

Another rule is to rebuild Qi after each treatment so as not to leave Qi reserves at a low level. Rebuilding Qi after each treatment ensures a practitioner can spend most of his life full of Qi. Many Qigong Masters have died young because they violated this basic rule. Some still feel that they are not truly serving unless they give all that they have. This attitude actually limits their ability to serve. A Master with weak Qi is far less effective at emitting and removing Qi than he would be with strong Qi.

In addition, his clarity of mind, and thus diagnostic ability, is depressed. Giving all he has on a routine basis means working far below his potential. Master Wan repeatedly emphasized the importance of doing Qigong exercises after treatment to rebuild Qi. This has major implications for all health care practitioners who experience burnout, pain, low energy, and fuzzy thinking.

Building and Moving Qi

Before we were ready to emit Qi, we had to have Qi to emit. Most people spend their lives with little available Qi inside them, so when they first try to emit, there is nothing to emit. This is why we spent so much time practicing Qigong exercises to collect Qi. We learned to collect energy from the sun, the moon, the trees, the earth, the water, and the air. The exercises varied considerably but all began with the same basic concepts:

1. Establish coherence, *Pei He,* with your surroundings, focusing clearly on the object with which you wish to exchange Qi.

2. As you inhale, draw Qi into your body by creating a vacuum in the lower abdomen. Start by using your imagination. With practice and increased sensitivity you will begin to feel the Qi flow. Then work toward increasing that sensation.

1. As you exhale, allow the impure Qi within you to flow back to the object with which you are exchanging Qi. Again, start with your imagination and become increasingly sensitive until you begin to feel the flow. Master Wan says that your toxic Qi is valuable to trees, water, and heavenly bodies. In the process of this Qi exchange, your Qi is strengthened and purified.

2. Become aware of how you are directing this flow of Qi. This is the most important part, for when you are consciously able to direct the flow of Qi, you become a Master of Qi.

Master Duan's Qi Field

The most effective exercise we learned for emitting Qi was taught not by Master Wan, but by a 91-year-old Chinese gentleman Master Wan introduced as *his* Master. His face was long and venerable. He had a wispy mustache and beard but his hair was almost pure black, though balding on top. His hands looked like those of a thirty-year-old. If Master Wan had accepted us as his disciples, this would have made Master Duan our Grand Master.

Listening to Master Duan speak, I was reminded of a Native American leader. His deep relationship with nature and his concern for man's lack of respect resonated with all of us. He encouraged us to spend time with nature, to know the Earth, and to exchange Qi with all living things. In this way we could feel the oneness with all of life, which was essential if we wish to make progress on the pathway of Qigong.

When a member of our group noted a cross hanging around Master Duan's neck and asked him about it, we learned that he was a Roman Catholic. His father, grandfather, and great grandfather had all been Catholic. He'd grown up in the "Forbidden City" and had been a playmate of Pu Yi, the Last Emperor. When drinks were served, he emitted Qi into them, and as he did, he looked just like a priest blessing the wine for Holy Communion.

Master Duan explained that Qigong is everywhere. It is an aspect of all religions and also an aspect of life where people hold to no religion. Although a Catholic was the last thing I expected, Master Duan had been a Catholic Qigong Master for 60 years, and it was perfectly normal to him.

He demonstrated many martial arts skills, such as having a strong young athlete try to knock him down, and each time easily guiding his attacker to the ground. One time his attacker knocked him over, and he gracefully rolled a reverse somersault over the cobblestones and sprang right back to his feet, showing the remarkable grace and agility that you would expect from an Olympic gymnast. A martial artist from our group also tried to spar with Master Duan, but always wound up on the ground.

Master Duan then divided us into two groups, separating us by about 30 yards across the courtyard. He asked us to hold up our hands and face the other group. He held up his hands as well. He then ran over to the other group and held up his hands facing us. I felt a strong sense of Qi flowing through my hands, as strong as I have ever felt emitted Qi. Most of the group felt this as well. Master Duan had created such a strong Qi field that we were able to exchange Qi with each other through a distance of 30 yards. We were amazed at this power.

How did Master Duan become such a powerful Master? He claimed that his power of emitted Qi came from doing the following simple exercise:

Master Duan's Qigong

Stand with your feet shoulder-width apart and legs slightly bent, with your eyes always following the Qi. Start, as in [Figure 1], with both hands in front of you. Then while inhaling draw your right hand toward your chest [Figure 2], pulling the energy of the Universe with you and drawing it into your lower Dan Tian. Then turn slowly [Figure 3] so that your hand is facing directly behind you and push while exhaling [Figure 4], feeling the Qi moving up with your breath from your Dan Tian and out through your hand, projecting it for miles into the distance.

Figure 1

Figure 2

Figure 3   *Figure 4*

Inhale again as you pull your hand toward your chest [Figure 4 again] and draw Qi powerfully into your Dan Tian. Exhale as you again push Qi out through your palms into the distance [Figure 2 again] and return to the starting position shown in Figure 1. Follow this pattern again with the left hand. The more slowly you can move and breathe, the more powerful the exercise will be.

Repeat this six times before moving on to the next movement.

Now move your hands out to the sides with palms facing outward as shown in Figure 5. Inhale and draw the Qi through your right palm as you pull your right hand in toward your chest [Figure 6].

Figure 5   *Figure 6*

Figure 7   Figure 8

Then exhale [Figures 7 and 8] as you push your Qi powerfully out
to your left through your palm, again using the energy from your Dan
Tian to propel the Qi along. Draw the Qi back again on your next
inhalation to be stored again in your Dan Tian [Figures 7, 6, and 5].
Repeat this exercise with the right hand. Then repeat this exercise with
both hands four times.

Repeat these first two steps completely three more times. The first
time you try this, your arms will soon feel like they weigh 100 pounds
each. With daily practice, not only will you gain physical strength, but
you will also gain the ability to move Qi. You will, in addition, learn *Pei
He,* coherence with nature, and strengthen your will. Most importantly,
you will learn to move Qi through your palms and to store it in your Dan
Tian. Spend a week practicing this exercise and you will begin to feel
strong Qi in your palms. Practice it for a decade and you will understand
Pei He. We spent just a few days with it before we were ready to try our
own hands at emitting Qi.

Richard Emits Qi

Despite all I had seen and experienced I still had my doubts about
emitted Qi, notwithstanding the strong evidence from China's scientific
laboratories and from Dr. Fu, who used Qigong in her practice. How
could I believe it when I could not emit it? And how could I believe that
I could emit Qi when I had no idea what emitting Qi was? I was clearly
at a disadvantage because of my scientific education.

At this point, back in Master Wan's emitted Qi training, we
separated into pairs, taking turns emitting Qi to each other. My partner

sat in a chair while I stood with my right hand above her head and my left hand extended away from her, shoulder high, with palm facing upward. I inhaled, drawing Qi through my hand and the top of my head into my lower abdomen. Then on exhalation, I guided the Qi up to my heart and out through my palm into my partner's head. After four or five tries, I started feeling the flow out through my palm! My partner said she felt it too, but she just might have been being polite, or suggestible.

Was it possible that I was emitting Qi? I looked around, wondering whether anyone else was experiencing this. I kept breathing and emitting! I knew how! It was just as Master Wan had described. And it was easy. Master Wan came over and watched as I tried to continue emitting Qi. I breathed and felt the flow of Qi. He squatted down and squinted at the space between my hand and my partner's head. I breathed even harder, though the flow felt about the same. Master Wan nodded his head and held his thumb and finger about ¼ inch apart, "Yi Dian" (A little bit)!

I had felt the flow and Master Wan had seen it. It was small, but it was emitted Qi! I felt very proud, even when Master Wan announced that other members of our group were emitting powerful Qi clear down into the chest of their partners. If I were a doctor, I could go back to my patients and try this when nobody was looking. As a scientist, I did not know quite what I would do with it.

The next day, we went off to walk on the Great Wall of China. We spent an hour on the wall, enjoying the scenery and snapping pictures, and another hour at the "trinkets for foreigners" area. As we waited for our bus, a member of our group came limping along with a sprained ankle. She stopped to rest, with her foot within my reach. I started the process of breathing and emitting Qi, aiming it into her swollen ankle for about 5 minutes. I could feel a flow similar to what I had felt the day before, but other than that, could see no changes. A few minutes later, however, when the time came to board the bus, the swelling was down and she walked the quarter-mile to the bus without pain. It appeared that I had performed Qigong healing!

Pulling Intoxication

Master Wan's group tended to serve alcohol at dinner and encouraged us to partake. One member of our delegation was a recovering alcoholic and did not appreciate the emphasis on alcohol. I asked one of Master Wan's students, a small, young, hundred-pound teenage woman, whether alcohol was an important part of her Qigong training. In our conversation, she told me that she could drink 22 shots of strong alcohol, often the champion, and that this was a sign that she was very good at controlling her Qi. This was most interesting to me. What does Qigong have to do with not getting plastered?

That night at dinner, Master Wan was busy pouring drinks. With each toast we were to empty our glasses, and he and the servers would immediately fill them up again. I was emptying my glass, and then quickly filling it up with a little bit of Coke and a lot of water to make it look like some sort of an alcoholic drink. Master Wan caught me. He invited me over and poured two glasses of alcohol as a demonstration. He then proceeded to hold his hand over one of them as if doing Qi emission, and with his fingertips, grasped the air above the glass pulled it away and released it as if he were tossing some bad Qi onto the floor. He then explained that he had removed the alcohol, or rather the intoxication quality from the drink, and encouraged me to taste the two glasses of wine. Sure enough, the drink he had worked on did not have a sense of alcohol whereas the other one did. I was fascinated.

Master Wan then poured two more glasses of wine and signaled me to give it a try. He wanted me to pull the intoxication qualities out of the wine. A part of me knew that I couldn't do that. But another part of me, which had just seen it done, and trusted that if Master Wan said I could do it, believed it was likely I would succeed. I went with it and held my hand over the wine, attempting to connect with the intoxicating qualities of the wine. I could actually feel the tingling that felt like intoxicating qualities. Amazing! So I closed my fingers around that intoxicating quality and pulled, just like Master Wan had done. It felt like pulling energy from a person during Qigong treatment. I could taste that the intoxication had been removed! Master Wan smiled his approval. I was amazed and proud.

After that experience, I wondered whether I still had the ability to pull the intoxicating qualities out of alcohol. I remember meeting Master Zhou in Los Angeles, where he demonstrated to us his ability to put aluminum foil on our legs, emit Chi into the aluminum foil, and have it get steaming hot. One of his students had asked him to teach this. Master Zhou refused. This was a precious secret. Then his student offered to pay him a lot of money to teach him, so he agreed. He taught the student to heat up the aluminum. But then the student didn't pay him the money, and he explained that the skill to heat up aluminum was withdrawn. The student no longer had that skill. For years, I wondered if I still had the skill of pulling intoxication out of wine. However, for some strange reason I never tried to find out.

One night I was out with some friends. One had too much to drink. He was feeling really intoxicated and sick. I thought back to my pulling intoxication experience and decided to try it. I imagined he was a wine glass, filled with intoxicating quality. I held my hand over his head, moving it around until I felt the intoxicating qualities as I had done many years before with Master Wan. I then grasped it and pulled it out and tossed on the floor. Sure enough, within a couple of minutes, my friend was feeling clear-headed and stable. That was amazing. I've tried it two or three times since, and it has worked reliably.

Not long ago while traveling in Belize, one of my co-travelers had received a whole lot of mosquito bites and complained that they were really itchy. She was very uncomfortable. I decided to try this technique and see whether the itchiness of a mosquito bite was similar to the intoxication quality of alcohol. I held my hand over her head again and felt. I found something that I thought might be a quality of itchiness. I grasped it as before, pulled it out and tossed it off the balcony. The itchiness disappeared from all but one of her mosquito bites. She felt it didn't reach that one because she was sitting on it. She remained free of it throughout the evening except for that one bite.

I wondered how far this could go. While studying awakening the third eye material in Australia (to be discussed in a later chapter), I had learned that astral materials are the cause of infectious diseases, or at least the fuel that causes them to proliferate rapidly. Over the last couple of years I wound up talking with two or three people who had colds and were miserable. I would ask if I could try to pull the cold out of their heads. They generally agreed with a strange, doubting smile, and invited me to proceed. I did the same thing, finding the tingling feel of it, pulling it out, and tossing it away. They all described substantial relief within 5 minutes. This was remarkable. It seems that vibration in the Qi is the cause, or at least the fuel, for many illnesses.

I was once visiting a friend who had a friend who had a complex infection she had caught in the hospital. She wasn't recovering from it. I volunteered to pull it out. She agreed and I proceeded to pull it out and toss it on the floor. While the sick friend seemed to get better quickly, a few hours later my friend was coming down with flu symptoms. Oops.

We used her pendulum skills and asked questions about this phenomenon. The problem was that I had poor housekeeping skills. For your reference, it is important to give the astral stuff you pull out someplace to go. You were probably thinking that all along. Why does Richard keep throwing that yucky stuff on the floor? The thought hadn't occurred to me. The recommended approach was to call in the help of someone like Archangel Michael to carry the discarded astral material to the light and dispose of it properly. An interesting and effective alternative, or additional technique, is to place the Scalene Light on the floor on Clear aiming up in the room, then to throw the extracted astral materials into the field above it. This appears effective at dissolving its infectious qualities. A small, wall-mounted Scalene Light is currently in development for use in hospitals, offices, retail stores, and homes to dissolve this substance in rooms and hallways.

It seems extremely important that medical science understand the importance of the Qi or astral component of disease. It seems that the astral component of healthy people is Qi, and that this stuff I was pulling out was the Qi of the microorganisms that fueled the disease processes. When hospitals learn to enhance the patient's Qi and dissolve the pernicious astral substances, it seems that there will be far fewer hospital-caused infections, and patients will come out of hospitals far healthier and more energized than they went in.

4. Infrasonic Simulation of Emitted Qi

"Until the 1980's, most Chinese considered Qigong Masters to be mythical storybook characters with superhuman powers."

I first met Professor Liu Guo-Long at the First International Congress of Qigong in 1990 in San Diego. As a medical doctor and neuroscientist who has conducted extensive research into the effects of emitted Qi on brain function, he had come to present some of his research.

Initial Skepticism of Qigong

"I was confident," Prof. Liu said looking intently at the group, "that my study of neurophysiology would prove that the emitted Qi was nothing more than a psychological factor induced by the waving of hands and hypnotic suggestion." He paused for a moment. "Instead I discovered that emitted Qi is very real!"

"Until the 1980s, most Chinese considered Qigong Masters to be mythical storybook characters with superhuman powers. However, when China's government began to support scientific research into Qigong, the few remaining true Qigong Masters surfaced, demonstrating such abilities as killing bacteria in test tubes, causing previously paralyzed people to walk, and lighting fluorescent tubes with energy emitted from their hands.

"I thought that these were all tricks of one sort or another, perhaps magicians' sleight-of-hand, hypnosis, or optical illusion. When I got the chance to work with people claiming to possess these abilities, I was fascinated. This was my chance to test, with scientific equipment, whether there was any truth behind the excitement about Qigong masters. I was confident that I could disprove the myths."

When I had first learned of these feats by Qigong masters, I myself, as a Westerner, also had doubts, so I was pleased that Professor Liu had been assigned the task of working with Qigong Masters who claimed to possess these abilities. This would allow him to test, with scientific equipment, whether there was any truth behind the excitement about Qigong Masters.

Professor Liu used the electroencephalograph (EEG) to analyze changes in the brain wave states of Qigong Masters. For several decades, EEG has proven a reliable means of mapping brain activity. Brain activity, what we call thinking, produces electrical impulses that are measurable at the surface of the skin. By attaching electrodes to several points on the surface of the head, scientists are able to determine what frequencies of electrical activity the brain is producing. Through a process called brain mapping, they can determine just exactly where in the brain these frequencies are occurring. This reveals whether any parts of the brain are not functioning normally, enabling scientists to pinpoint where a disease process may be active or where abnormal brain activity might be causing other problems.

Medical Infrasonics

Professor Liu continued, "Our research grants were motivated by the pioneering work of Senior Engineer Lu Yanfang of the National Institute of TV and Electro-Acoustics, who developed the Infratonic device, a device that simulates the infrasonic energy output from the hands of Qigong doctors."

"Engineer Lu had determined that humans have a very high degree of acoustical activity in the subsonic range below 20 Hertz, similar to the alpha rhythm of the EEG. The alpha rhythm is a pattern of coherent activity between 8 and 14 Hertz throughout the brain. Engineer Lu also found that people with chronic illnesses had a much lower level of infrasonic activity than those who were healthy, and that Qigong masters had a much higher level of infrasonic output than normal when they were emitting Qi.

"The implication," Professor Liu continued, "is that infrasound might be related to human functioning and, further, that it might be involved in the mechanism of brain functioning. We searched the available scientific literature for research papers clarifying the relationships between sound waves and brain waves, but found nothing.

"The only research we could find was that done with the *Infratonic*. *[The term he used to describe the Qigong Information Instrument]* Thousands of hospital patients in Beijing had been effectively treated with it for a wide variety of conditions ranging from painful joints to digestive problems, from uterine bleeding to surgical complications. Laboratory tests had verified the effectiveness of this machine."

Here's his paper:

Infrasonic Simulation of Emitted Qi

By Professor Liu Guo-Long, M.D., PhD
Beijing College of Traditional Chinese Medicine

Summary of Research

Emitted Qi from Qigong Masters clearly has a strong effect on the central nervous system, not only in humans but also in animal subjects. Qigong meditation and the Infratonic have similar effects. These findings are summarized below:

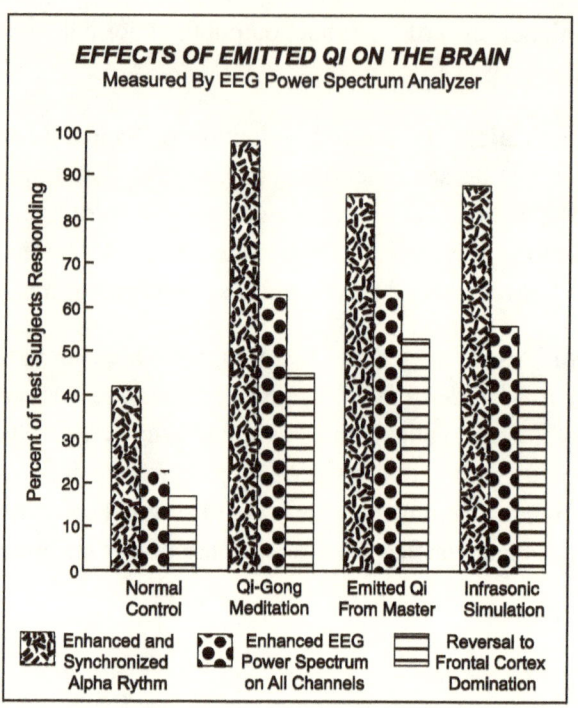

Emitted Qi has a pronounced and repeatable effect on EEG. It enhances frontal and occipital EEG power spectra, and often enhances the frontal so much that frontal becomes the dominant EEG activity, whereas occipital dominance is more common. Emitted Qi also enhances and synchronizes the Alpha.

Background

I was a specialist in Western medical technology assigned to the research department of the Beijing College of Traditional Chinese Medicine. The project was sponsored by the Chinese Government Department of Education and the Department of Natural Sciences. My research group was assigned to study the relationship of infrasonic waves to emitted Qi. These research grants were motivated by the pioneering research of Prof. Lu Yan Fang at the National Institute of Electro-Acoustics in Beijing, who developed prototype devices to simulate the output of Qigong doctors.

Research has confirmed that humans have a very high degree of acoustic activity in the subsonic range below 20 Hz (infrasonic), similar to the alpha rhythm of EEG. Also, people with chronic illnesses were found to have a much lower level of infrasonic activity, while Qigong Masters had a much higher level of infrasonic output when emitting Qi.

The implication was that infrasonic sound might be related to human functioning, and further, that it might be in some way involved in the mechanism of brain functioning. We searched the available scientific literature for research papers that described relationships between sound waves and brain waves, but found none that had been done.

Extensive clinical research, based on infrasonic Qigong simulation, showed it to be effective for a wide variety of hospital problems. These results are what motivated the National Department of Education and Natural Sciences to look further into the importance of infrasonic sound and what led to the research I am about to describe:

The Brain as a Detector of Emitted Qi

I had learned in my many years of research with the Electroencephalograph (EEG) that the human brain responds to even the most subtle of stimuli to the body, so I reasoned that, if there were really any scientific basis to emitted Qi, it would show up in the brain waves of test subjects who were placed in the path of these emissions. I expected to see no difference between the resting states and the Qi emission states.

What we saw was extraordinary. Within a few seconds after the Qigong Master began to emit Qi, the subject's EEG would begin to shift. The EEG power spectrum was enhanced in all channels while the most pronounced increase was in the frontal lobe. Also, there was an enhancement and synchronization of the Alpha Rhythm in all channels. When the Qigong Master stopped emitting Qi the EEG would gradually shift back toward the baseline readings.

To determine whether infrasonic energy was a significant part of the emitted Qi, we used the infrasonic Qigong prototype in the same experiment. It was located 18 inches away, directly behind the back of the head of the test subject. The EEG electrodes were attached as before. The simulator was activated for short periods of time and the results recorded. We found that the effects on the receiver's EEG were quite similar to those of the emitted Qi.

Our further research involved monitoring the various sensory-cortical evoked potentials during Qigong meditation, emitted Qi, and infrasonic Qigong simulation. We again found very similar results from all three stimuli. We found that a large portion of the cerebral cortex was inhibited while other areas of the somatosensory cortex were excited. One of the significant findings of this study is that *the inhibition of the cerebral cortex during Qigong meditation is clearly different from the excitation of the cerebral cortex that is measurable during sleep.*

Through Acoustical Brainstem Evoked Response (ABER) it was found that the brainstem structures from the medulla to the hypothalamus were significantly facilitated. The brainstem plays an important role in regulating the functions of the inner organs, motor function, and emotion.

The implications of these studies were startling. Qigong Masters can, without touch, voice, eye contact or any other traditional means of communication, induce a clear, strong, and highly measurable change in a subject's brain functioning. A synchronization of alpha rhythm indicates deep relaxation, and is closely associated with accelerated healing. Enhanced power spectrum in the frontal lobe is especially significant because the association cortex of the frontal lobe is concerned with higher motor action, higher sensory function, emotional and motivational aspects of behavior, and integration of autonomic function. Facilitation of the brain stem, with its regulation of internal organs, may be a mechanism by which physical healing is induced or accelerated.

Despite these highly significant changes in EEG and evoked potentials, the subjects had felt nothing and had no idea of the profound changes taking place within them.

The findings of these studies are solid evidence that a Qigong Master can induce real physiological changes in a subject from several feet away, and further, may help to explain the high rate of recovery from chronic degenerative diseases in groups of hospital patients under the care of Qigong Masters. These studies also show that the infrasonic Qigong simulator can induce similar changes in brain function and that, through Qigong meditation, a Qigong Master can induce these same changes in his own brain.

Scientific Control

There is much disagreement on how emitted Qi affects the brain. Many doctors insist that brain changes are psychologically induced, and that verbal suggestion, impressive hand motion, and expectation of the subject account for the observed phenomena.

To test this, we had several Qigong Masters and people pretending to be Qigong Masters treat the test subjects. The subjects were told that all were Qigong doctors, and all moved their hands in similar ways. We saw no significant changes in brain wave patterns with the fake Qigong Masters, but when the real doctors emitted their Qi, or when we used the infrasonic Qigong simulator, we repeatedly got the highly significant changes.

Even this did not satisfy many of the doctors who reviewed our work, so we repeated the study with animals. We monitored EEG in awake rabbits and ABER (Acoustical Brainstem Evoked Response) in anesthetized cats as Qigong Masters emitted Qi toward them. Even though there was no voice or eye contact between the Qigong Masters and the animals, and the Masters emitted Qi from several feet away, we saw shifts in EEG and ABER similar to those observed in the human subjects. This is a highly convincing result because all kinds of placebo effects are eliminated, yet modification of brain function at a distance remains.

Infrasonic vs. Electromagnetic Interference

Extremely low frequency (ELF) electromagnetic signals can affect BEG (ELF modulated EEG readings). There is growing concern about the low-frequency radiation produced by 50 and 60Hz electrical power lines located close to people's homes and schools because of apparent disruption of brain and cellular function. ELF signals will cause the principal BEG power spectrum to show a spike at the frequency of the ELF signal. This is because the EEG is easily entrained by ELF signals. When the signal is discontinued, the BEG abruptly returns to normal. Electrical power lines operate at 50 or 60Hz, which corresponds to BEG in the high Beta range, associated with mental anxiety and confusion. The entrainment of EEG at 50/60Hz around power transmission lines may be why researchers are finding that people who live close to these power lines show a higher incidence of brain tumors.

The results using the infrasonic Qigong simulator were quite different from ELF signal entrainment. The shifts in EEG were gradual rather than abrupt, and while dominant EEG frequency did drift toward the dominant peak infrasonic frequency, it was a broad spectrum of EEG activity rather than a spike. The enhanced power spectrum continued after simulation was stopped, gradually decreasing and returning to the pre-test state. From these observations it is clear that the effects of the infrasonic simulator are quite different from the BEG entrainment of ELF electromagnetic signals.

Summary

When I started this project in 1976, I had serious doubts about Qigong Masters and emitted Qi. Now I am convinced that emitted Qi is very real and that Qigong is a very valuable art. With the support of the China government, an estimated 50 million Chinese are practicing Qigong daily, and many Western hospitals have added Qigong departments for patients with chronic and degenerative diseases. In addition, the Chinese government has funded extensive scientific research into the nature of emitted Qi with the goal of advancing science and medical technology.

The study of emitted Qi and infrasonic sound, as they relate to human health and functioning, is a broad and exciting field. The use of Qigong in treating chronic degenerative diseases such as cancer and hypertension in China has proven very effective. It has been employed to accelerate healing for thousands of patients with a wide variety of diseases in Chinese hospitals. I am confident that emitted Qi, Qigong meditation, and the infrasonic Qigong simulator will play an increasing role in health care around the world.

5. Quantifying Infratonic Therapy Personally

It was certainly wonderful to have the broad spectrum of Chinese research showing the effectiveness of, and providing insight into the mechanism of, Infratonic therapy. However, I always insist on personal verification to believe. I needed to verify its effectiveness personally to the extent I could. Thus I set out to conduct as many experiments as possible to see for myself the effectiveness of the Infratonic, and maybe to uncover more of its mechanisms. This chapter is devoted to some of these studies.

Electromyogram Studies

In an early experiment I arranged to have Brian Gwartz, M.D. examine the effectiveness of the Infratonic on chronic back pain using electromyogram (EMG). EMG electrodes were attached to the backs of test subjects with chronic back pain. Subjects were then instructed to bend over repeatedly for a period of 200 seconds. The EMG measures the voltage spikes produced by the back muscles as they are exerted. The first graph, below, shows the EMG of a person with minimal back pain who bends over easily:

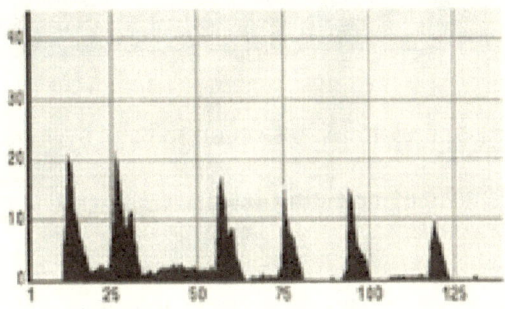

Four observations can be made regarding this patient from the EMG:

The subject bends over and back quickly, in about 5 seconds.

The voltage produced is low, about 15 microvolts (μV) and decreases with repetition.

Of greater importance is that the curves are relatively smooth, indicating that the muscles respond quickly and easily to the request.

The baseline voltage between the peaks is low, about 10 to 20μV. This means that when the muscles are not moving, they are quiet. They are not producing voltage. This indicates again that the muscles are responding well to the brain.

This second graph shows a patient who has chronic back pain and spasms, and is in considerable pain. This EMG shows a patient with chronic back pain, and is taken before any treatment is administered. Again, the patient bends over repeatedly.

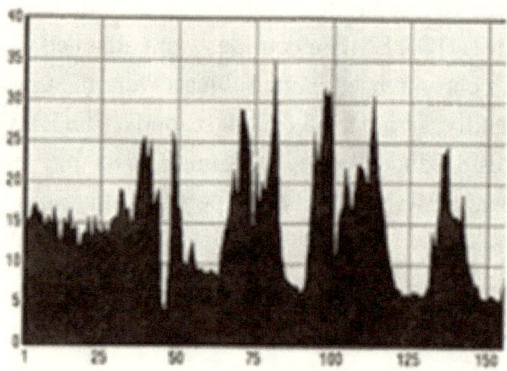

The subject requires 15 to 30 seconds to bend over and return.

The voltage is high, indicating extensive electrical activity in the muscles.

The curve is erratic during the process, indicating that the muscle is still highly electrically active during rest, which indicates continued pain.

The baseline voltage is 6 to 8μV, which shows that even during rest the muscles are not resting, indicating residual or resting discomfort.

The lower back of the test subject was then treated for 15 minutes with the Infratonic. Then the subject was scanned again. This third graph shows the following changes in muscle activity, indicating reduction in pain:

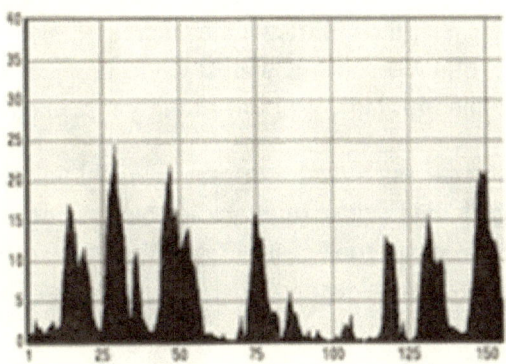

The subject bent over and returned more quickly, indicating greater ease of motion.

The peak voltage dropped considerably, showing reduced pain during movement, though voltage remains above normal.

The curve is much smoother, indicating fewer and less intense points of pain were encountered during the bending process.

Baseline voltage between periods of muscle activity decreased from about 8 micro volts to about 2μV, indicating reduced pain during rest.

This study supported patients' experiences of reduced back pain with use of the Infratonic.

Neurofeedback Research

I next traveled to the Washington DC area to visit Dale Patterson, a neurofeedback specialist who was interested in researching the

Infratonic. We tested several test subjects, applying the Infratonic to the center of the chest, and found that, consistent with Professor Liu Guo Long's research, the Infratonic did induce increased alpha activity in the brain. It provides similar increase in alpha brainwave activity when applied to the bottoms of the feet.

In one very interesting finding, we found one test subject who said he didn't really like the Infratonic. Surprisingly, his alpha activity decreased during use of the Infratonic, then increased dramatically once therapy was discontinued, showing a most interesting resistance to therapy, and that the Infratonic was still strongly influencing alpha.

We had just come out with our Infratonic signal with "Chaos Therapy" that had a high degree of unpredictable variation in frequency and amplitude, so it was a good chance for us to test the difference in effectiveness between the older and newer technologies. The following graph shows these results:

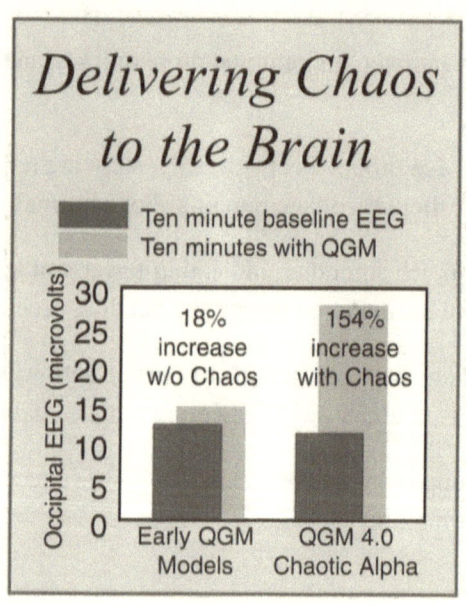

In this study of changes in EEG brainwave activity it was found that the power of alpha brainwave activity increased far more dramatically with the application of the highly unpredictable Infratonic, as opposed to the older model. Why was this?

The Healing Benefits of "Pink Noise"

Research with the Infratonic in China had all been based around infrasonic sound. Because infrasonic sound was measured, it was assumed that infrasonic sound was the effective ingredient. When Maria Gonzalez, PhD, a research scientist in the area of cellular and genetic biology, joined my company and started researching infrasonic sound, all she found in the literature were studies that showed that high-amplitude infrasonic sounds are damaging to our health. Jet planes, traffic noise, loud rock concerts. All are detrimental to health. Her research into infrasonic sound also showed that the military, both in the United States and in Russia, had invested extensively in the study of infrasonic sound as a weapon. Thus, as Dr. Gonzalez surveyed the literature she found very little to support the health claims of the Infratonic.

She dug further and found two other categories of research that were far more fruitful. Extensive research had been conducted with what was called high-frequency mechanical stimulation, which showed effectiveness in a variety of applications in the general frequency range of the Infratonic. She also found a most interesting field of research, "stochastic resonance therapy." This seemed to describe, and provide a theoretical framework for, many of the benefits observed with the Infratonic. In probability theory, a *stochastic* system is one whose state is non-deterministic. Stochastic resonance therapy utilizes random noise which is sound with probabilistic uncertainty as to its frequency, but within specific frequency limits.

Stochastic Resonance Therapy and the Infratonic

From one perspective, the phenomenon of stochastic resonance is defined as an increase in the quality of signal communication rather than a decline as a result of increased random noise introduced into a system. A broader definition of stochastic resonance covers all activity in which random activity improves signal processing. Random noise, such as that produced by the Infratonic, influences cellular and genetic structures as a result of inducing stochastic resonance signals into the body.

Computerized studies of multi-variable systems have shown that single-frequency signals tend to destabilize these systems, and amazingly, random signals tend to stabilize them. This is perhaps most easily illustrated with music. A very friendly oboe or flute playing a single note continuously becomes tedious and annoying, even stressful. Contrast this to the same instrument playing music consisting of constantly changing and seldom repeating melodies. This tends to be highly relaxing and therapeutic. If you want an entry into the literature, look up "Biological Functions of low-frequency Vibrations (Phonons)" by Kuo-Chen Chou, published in the *Biophysics Journal.*

The paradoxical concept of "good noise" is not particularly appealing to researchers. There is a prevailing belief, often justified, that noise interferes with communication. However, chaos theory tells us that when we increase the energy of a system it can tend to reorganize into a higher, simpler state of organization. Stochastic resonance research probes the systems of the physical body in an attempt to discover those systems which are subject to and can be positively influenced by noise.

Surprisingly, noise seems to be central to the functioning of the nervous system. Neurons typically tend to fire at a frequency between five and 20Hz. After they fire they need a rest-and-recovery period to rebuild their energy. It takes about this long for them to fire again. When they actually fire is determined to a large extent by noise, by all the signals of the body, particularly the signals of hundreds or thousands of other neurons that are firing away, introducing additional energy to the body in a form that neurons need. This allows for an increase in frequency of firing. Other systems in the body also function partially on the basis of noise, including molecular interactions and stochastic gene expression.

Stochastic resonance first came under study as a concept to explain periodic behavior of Earth's ice ages. It got a substantial boost between 1989 (the year after the Infratonic was first introduced into the US and world market) and 1992, when about 15 papers were published per year. The next significant increase in stochastic resonance research came between 1993 and 1997 when several papers were published on the physiological measurements of neurons. 1993 to 1997 was also when I took six separate groups of doctors and researchers to China to visit Master Wan, and observe his emitted Qi methods. I advertised Master Wan's treatment methods and graphs similar to those shown below during this period with more than a million mailers to the alternative medicine community. Around 1997 the number of papers started increasing rapidly. This was the year that my company introduced the first-ever highly unpredictable mechanical stochastic resonance device to the world market with the trademark Chaos Therapy. Between 1997 and 2002, the number of published papers increased to over 200 per year and has continued at roughly that rate since then.

Does the Infratonic improve signal-to-noise ratio in the body? Our research with the trembling analyzer shows that it does. This research shows that trembling in the body is substantially altered when the Infratonic is applied. Noise is substantially reduced and becomes more focused in the EEG range of alpha. The image below shows an output of the trembling analyzer while measuring me trembling with a fair amount of beta activity.

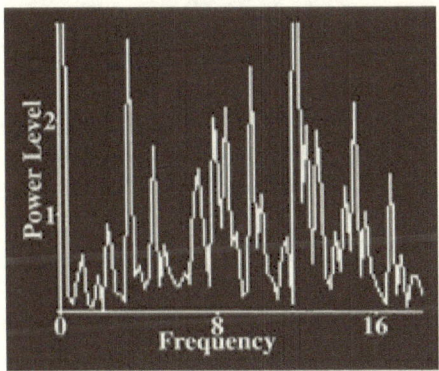

Notice the high trembling activity in my hand from 12 to 18 Hz, and the overall high activity.

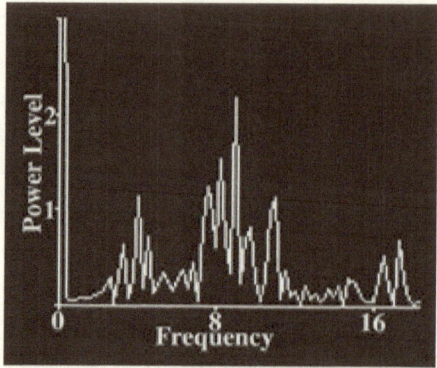

The above image shows my hand trembling during the application of the Infratonic after about 10 minutes. Notice the decrease in overall trembling and the shift toward the 8 to 12 Hz, alpha range.

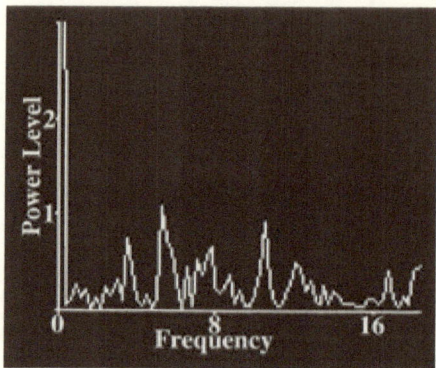

Finally, this image shows my trembling 10 minutes after stopping the Infratonic therapy. My beta activity is way down. I was feeling far more relaxed. Compare to the first graph and you will see that my overall trembling at all frequencies has decreased. As this noise in my body decreases, I am more sensitive to and aware of my surroundings, more able to think and perceive clearly.

Noise is important to the functioning of the systems in our bodies, not just the nervous system but many systems. Since noise is shown to affect performance of these systems, it is important to understand how different qualities and quantities of noise affect these systems. Thus, strategic forms of noise can be important pathways to enhancing human performance and providing optimal health.

It was wonderful to be able to personally validate the laboratory research conducted in China. However, I wanted real clinical verification. Unfortunately it is very difficult to research the treatment of illness in people because everybody is different and every condition is different. Clinical research would require large population samples and require far more money that I had available. Besides if scientists want to know whether something really works, they use animals that can be fed the same diet, have the same condition needing treatment, and do the same activities. As it turns out, horses were perfect.

Controlled Equine Research

In the year 2001 a thoroughbred stallion named Monarchos won the Kentucky Derby in the second-fastest time in the history of the race. He had used our Infratonic (Equitonic version) regularly to relax and to recover from strenuous workouts. Every year, many of the best horses don't race because they are injured during the grueling season and unable to race at the end of the season. Monarchos remained healthy. Since then, 4 or 5 horses every year that used the Infratonic have remained healthy enough through the season to race in the Kentucky Derby.

Because of the remarkable success achieved in treating injured horses and having many of them come back from injuries to win prestigious races, I began a series of research experiments to explore the mechanisms behind its actions, conducted by Ronald J. Riegel, a veterinarian. These studies reveal valuable insights into to the mechanisms behind inflammation, which can contribute to accelerated recovery from workouts, reduced injuries, and faster recovery from injuries. They also get around the main complaint against most clinical research, that it is subjective. My aim was to conduct controlled animal studies that would be solid and repeatable. Toward this end, Dr. Riegel utilized groups of 10 standardbred racehorses, all genetically similar, eating the same diet, and participating in the same daily workout schedules. Thus standardbred horses became an amazing vehicle for quantifying the effects of the Infratonic.

Our first study utilized infrared thermography to compare the left and right hocks of 10 standardbred racehorses. The thermographs below show the reduction in inflammation when just one hock was treated with the Infratonic for 20 minutes:

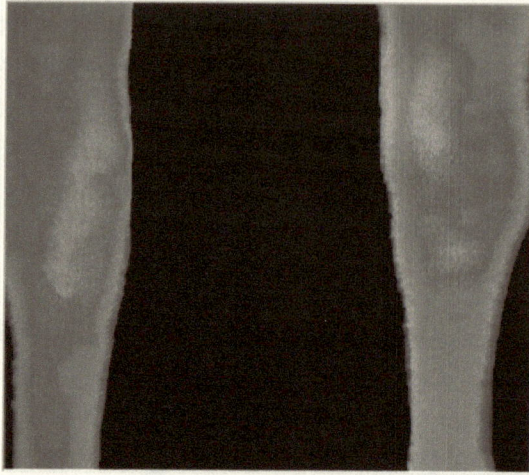

Initial Thermogram shows inflammation (orange and yellow) in both hocks of a typical standardbred racehorse before Infratonic treatment.

Treated . . . . Control

30 minutes after QGM treatment of 12 acupoints, the treated hock dropped more than 3 degrees Celsius (the black indicates temperature dropped off the scale).

Treated . . . . Control

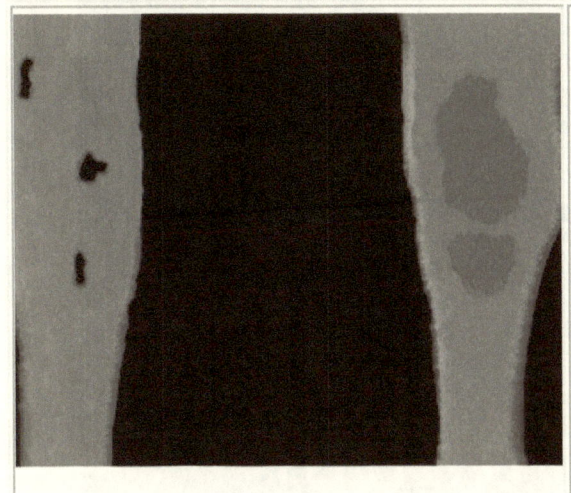

72 hours after treatment the 3 degree Celsius drop remains, and the control hock also shows a decrease. Trainers reported greater strength in the treated side of all 10 horses.

Treated . . . . *Control*

After only one treatment of 40-45 minutes with the animals still in training, substantial decreases in thermal gradients were observed, which did not start to return to their pre-treatment state until 96 hours post-treatment in all ten animals. The next question was "What happens with multiple treatments?" Often effectiveness of a therapy will decline with repeated application. In this case we saw the opposite.

The following chart summarizes the results of three studies conducted with these groups of standardbred horses a week apart. These findings show progressively more effective reduction in inflammation, which indicates that repeated treatments over days or weeks offer a cumulative effectiveness far greater than individual treatments alone. On the first week, a 10-minute treatment was administered. The second week, a 20-minute treatment, and finally, on the third week two 20-minute treatments were given, 20 minutes apart.

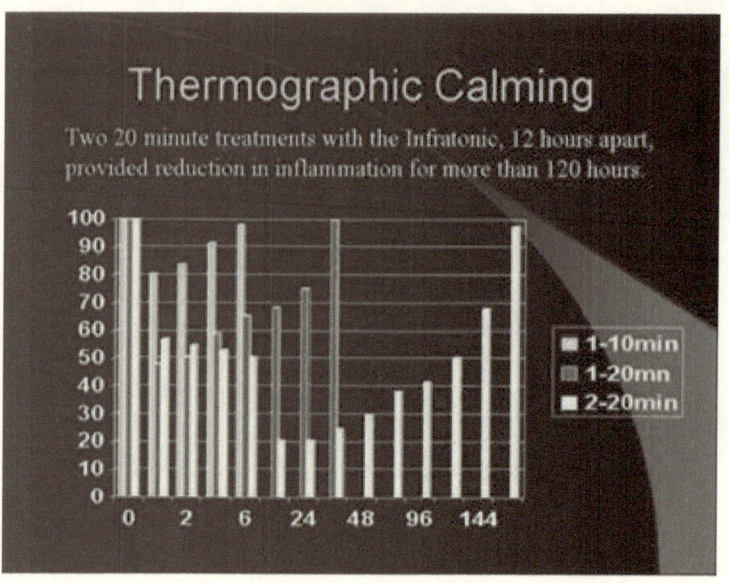

Thermographic Calming

Two 20 minute treatments with the Infratonic, 12 hours apart, provided reduction in inflammation for more than 120 hours.

Note first that the first treatment on the third week provided better results than that on the second week. (Compare dark blue and yellow at 6 hours). Also note that a second 20-minute treatment, 12 hours after the first treatment, provided profoundly deeper and longer-lasting relief in inflammation, indicating that Infratonic therapy is cumulative, with two or more treatments providing substantially more benefit than a single treatment.

When I saw these findings I thought they were quite impressive in that the Infratonic was shown to have such a profound effect on inflammation. However, inflammation, as measured by infrared thermography, is not universally accepted because it is simply a measure of increased skin temperature, which while clearly indicating that inflammation is present, and in this case has been reduced, offers no indication of mechanism, and does not incorporate the "gold standard" of clinical research, measurable changes in blood chemistry.

The Gold Standard: Bloodwork

I initiated a second field trial, utilizing two groups of ten standardbreds in active training. This time, it included blood drawn from the horses, which revealed that Infratonic therapy brought about a significant drop in both the AST (Aspartate Amino Transferase) and CPK (Creatine Phosphokinase) levels within the blood serum in all ten horses in the treated group, and an increase in these measures in the control group. Both of these enzymes are indicators of inflammation within the muscle tissue. Their reduction reveals a healing or anti-inflammatory effect taking place within these tissues. As before, all horses were undergoing daily workouts, and thus the worsening condition in the control group is due to the regular workouts. Further, the improvement in the treated group is remarkable because, in spite of the continued daily heavy workouts, the inflammation, as measured by AST and CPK, decreased. The blood evaluations did not reveal any negative side effects from this treatment.

These graphs summarize the changes in concentration of the diagnostic enzymes CPK and AST in groups of 10 standardbred horses over six weeks treated 45 minutes per day:

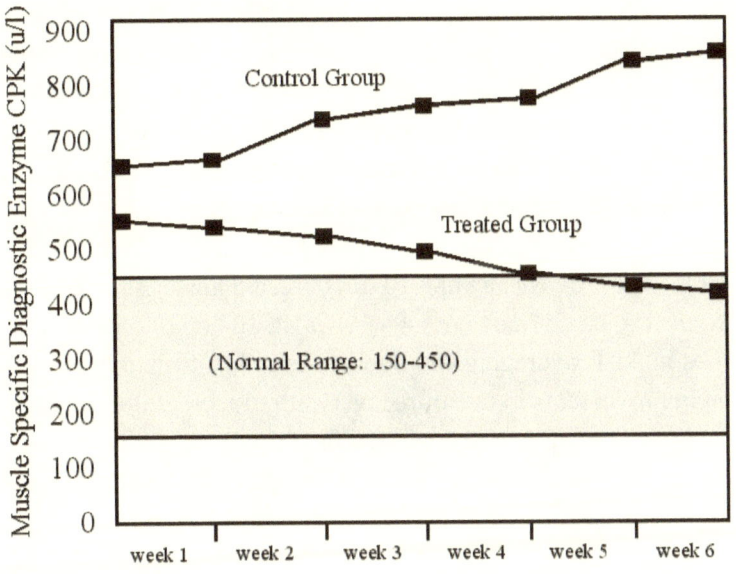

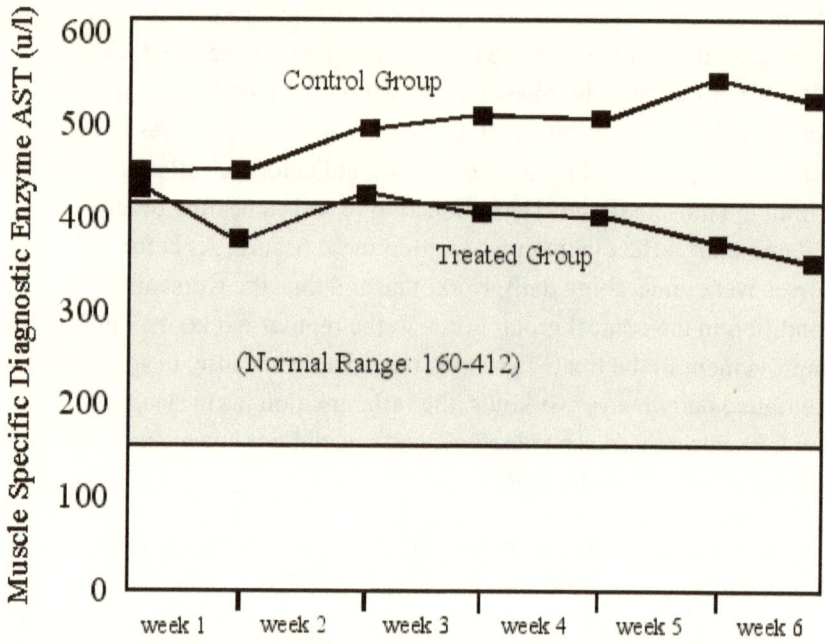

This study utilizing standardbred horses was highly controlled, with all horses eating the same food, utilizing the same workouts day after day, and cared for by the same trainers. Except for CPK and AST, all horses had normal bloodwork throughout the study. Nine of the ten control group horses showed worsening race times and bloodwork, and 4 of the 10 became lame and dropped out partway through the study. Average CPK levels for the control group went up from 665 to 857 (normal is 150-450). Meanwhile 10 of 10 treated horses showed decrease in CPK, averaging 557 down to 404. Also, all 10 horses showed decreases in AST, averaging 444 down to 352. The treated horses also showed improvements in rate of recovery after races, attitude, and performance, shaving 1.65 seconds off their time over six weeks, enough to make the difference between last and first place in many close races.

Thus it comes as no surprise that so many horses treated with the Infratonic remain healthy and are able to compete in the Kentucky Derby.

Increasing Hyaluronic Acid (HA)

I was very pleased with this research with standardbred horses. Still, a question lingered in my mind. Can we identify the biochemical mechanism by which the different measurables could all be explained? The infrared thermography showed us that inflammation was really decreasing. The reduction in inflammatory chemicals showed us that damage to muscle cells was being repaired, and less of the diagnostic enzymes CPK and AST were leaking through the cell membrane of muscle cells into the blood. Yet I felt that a mystery remained.

Research told me that hyaluronic acid, a key ingredient in skin and muscles, joints and eyes, was very important to maintain health and promote healing. I discussed this with Dr. Riegel and he explained that veterinarians had been injecting hyaluronic acid into the "dry" hocks of horses for many years, and that it was highly effective at reducing inflammation and improving the performance of horses. Inflamed hocks are similar to the knees and hips of people with chronic inflammation. "Dry hocks" in horses is parallel to what is often referred to as "bone on bone," by people with painful joints. While injection with hyaluronic acid was illegal 12 years ago, it is now common for people suffering from chronic joint pain.

Might Infratonic therapy be increasing hyaluronic acid concentration in the muscles? Might this be increasing the coherence of the cell membrane and reducing the leaking of CPK and AST from inside the muscle cells into the bloodstream? Dr. Riegel felt that it would be difficult to measure the hyaluronic acid concentration in muscles or in the blood, but easy to measure it in hocks, where we had measured the reduction in inflammation with infrared thermography. Thus the research question became, "Will Infratonic therapy increase hyaluronic acid concentration in the hocks of horses?"

What we found was dramatic. After 5 weeks the hocks treated with Infratonic therapy had 40% more hyaluronic acid than untreated hocks. This research shows that joints which are inflamed and "dry" (containing a reduced amount of synovial fluid) are rejuvenated after regular Infratonic treatment.

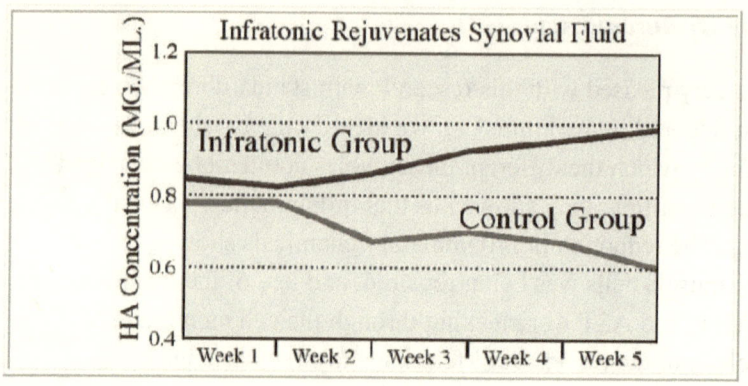

Ten standardbred horses under rigorous training showed a significant increase in Hyaluronic Acid (HA) from use of the Infratonic, while the control group showed a decline.

After just 5 weeks of treatment, we found not only that the concentration of HA had increased, but also that the quantity of synovial fluid had increased dramatically so that the joint was now fully lubricated and pain and inflammation had all but disappeared in all treated subjects. In addition, the quality of hyaluronic acid improved. Since starting to play volleyball for two hours per night, three times per week, I can attest to the value of Infratonic therapy for reducing pain and inflammation in my knees and shoulders.

The implications are clear. Inflammation damages or destroys hyaluronic acid and makes joints "dry" and painful. Infratonic therapy reduces this inflammation so that the synovial membrane, instead of producing inflammatory chemicals, produces hyaluronic acid. The same thing happens in muscles. The inflamed cells stop producing inflammatory chemicals and start producing more hyaluronic acid. Thus hyaluronic acid explains how the Infratonic can quickly reduce pain and at the same time accelerate recovery.

Hyaluronic Acid and Rejuvenation

As I researched hyaluronic acid, I learned that, according to the literature, the body creates about 5g of it each day, and about 5g is destroyed each day, mostly in the skin and intestines. It appears that hyaluronic acid is destroyed partly because it is one of the first lines of defense against high-energy free radicals that penetrate the body through the skin. Thus free radicals and inflammatory chemicals produce the same corrosive effect, damaging surrounding cells and destroying hyaluronic acid. Also, hyaluronic acid in the stomach and intestines is damaged by the free radicals contained in the food we eat. As free radicals destroy hyaluronic acid, they are neutralized. Thus hyaluronic acid behaves as an antioxidant, protecting internal organs from damage from multiple reactive species. Small children, who recover quickly from injury, have a very high concentration of hyaluronic acid, whereas adults have a decreasing concentration. Thus hyaluronic acid has the potential for being a valuable partner in rejuvenation.

Hyaluronic acid appears to be so important to the body that I started a little project to see what other factors might influence the concentration of hyaluronic acid in the body. While I have not done the controlled research to prove this, it is my belief based on experimentation with my body that a diet rich in antioxidants increases hyaluronic acid by protecting the digestive system. This can be seen in softer skin, faster healing, and reduced inflammation. Key considerations: Processed oils, particularly frying oils, are highly reactive, essentially pouring free radicals into the digestive system. Eliminating processed and fried oils dramatically increases hyaluronic acid concentration. This diet also eliminates most processed foods in the middle aisles of supermarkets.

Potassium Citrate is a powerful antioxidant. Most people are deficient in dietary potassium. I started adding 1 teaspoon into my breakfast each morning. The potassium citrate alkalizes and buffers the blood against free radicals, substantially reducing the load on the hyaluronic acid.

One teaspoon provides about 3g of potassium, half of the minimum daily allowance. A larger amount is not recommended because there are certain pharmaceuticals that block the elimination of potassium, and high doses of potassium, if combined with these pharmaceuticals, can cause serious health problems. *Note:* Restrictive regulation of potassium supplements means that you can't buy them with more than 99mg of potassium. It would take 20 pills to equal one teaspoon of potassium citrate powder. A pound of crystalline powder cost me about $15 on the web, and is equal to about 5,000 of those little pills.

Finally the biggest loss of hyaluronic acid comes through the skin as a result of ionizing ultraviolet radiation, or the resulting airborne free radical, hydroxyl, OH-. This energy ionizes molecules within the body, creating free radicals. I've been experimenting with different antioxidant-loaded lotions applied to the skin. It appears that antioxidant lotions protect the skin from airborne hydroxyl. They appear to make a significant difference in reducing the amount of free radicals that enter through the skin, thus reducing the load on hyaluronic acid, and contributing to accelerated healing, reduced pain and inflammation, and rejuvenation.

To illustrate the importance of protecting this skin barrier to protect against free radicals, a study at Jikei University in Tokyo found that cardiopulmonary bypass that occurs during cardiovascular surgery (which bypasses the protection of the skin by removing the blood from the body) leads to the production of free radicals and oxidative stress. According to this study, this does not happen with procedures of similar duration involving an aortic clamp. I would attribute this to taking the blood out of the body and running it through a machine with lots of steel, in a room with fluorescent lights that produce high-energy ultraviolet ionizing radiation which goes right through steel. Thus heart bypass machines cause substantial oxidative damage by removing the blood from the protective skin of the body and allowing it to be filled with oxidizing radiation. By analogy, having skin with a low level of antioxidants also allows oxidative damage, and damage to the hyaluronic acid.

To summarize: The first conclusion is that sound, unpredictable sound in the range of the alpha rhythm, can have a profound physiological effect on traumatized tissue, relieving pain and inflammation and accelerating recovery. The second conclusion is that hyaluronic acid is a very important building block of the body, which is damaged by processed and cooked oils, and particularly by fried foods. Eliminating processed and fried oils from your diet, adding a little potassium citrate, and applying an antioxidant body lotion regularly can make a huge difference in the level of hyaluronic acid in your body and the quality of your life.

Diabetic Neuropathy Research

Finally, I would like to share with you some of our latest research, and what led up to it. We had been receiving calls from people with symptoms of diabetic neuropathy, claiming that their pain, discoloration, ulcers, and other symptoms were diminishing from the use of the Infratonic. This was amazing to me because diabetic symptoms are generally considered to be progressive and irreversible. I decided to conduct a study with patients of doctors who were on our mailing list. We enrolled 20 subjects who had not used the Infratonic, and had their doctors fill out questionnaires about their condition before and after the test. We also provided an Infratonic and instructed the subjects on the protocol. Seventeen of the 20 made it through the first 2 months.

When we did this study, 6% of Americans had diabetes, and half of them, or 8.5 million Americans, suffered from some form of neuropathy, according to the American Diabetes Association. The number has risen over the last 10 years to 25.8 million. Not only is this the fifth-deadliest disease, contributing to hundreds of thousands of American deaths annually, but also the cost of health care and lost productivity due to diabetes is in the trillions of dollars annually. The peripheral complications associated with diabetic neuropathy, like pain, disability, vascular disease, and nerve degeneration are considered progressive and irreversible. The prognosis is generally progression of the disease, ongoing pain, amputation of digits or limbs, and increased disability. Improvement is considered unlikely.

Thus, it is astounding that our group of 17 study participants, each with advanced diabetic neuropathy symptoms, showed improvement in every measure of this study. While these findings are far from conclusive, they are promising.

The Diabetic Neuropathy Protocol

We solicited 20 diabetic patients from doctors on our mailing list, and entered them into the study based on both the existence of substantial symptoms and the completion of an examination and report from their doctor. We then had them treat the bottoms of their feet for 10 minutes each, then the belly over the stomach and pancreas, and the thymus along the upper sternum. By the end of the two months, 17 of the 20 patients were still engaged in regular therapy. The findings for these 17 are as follows:

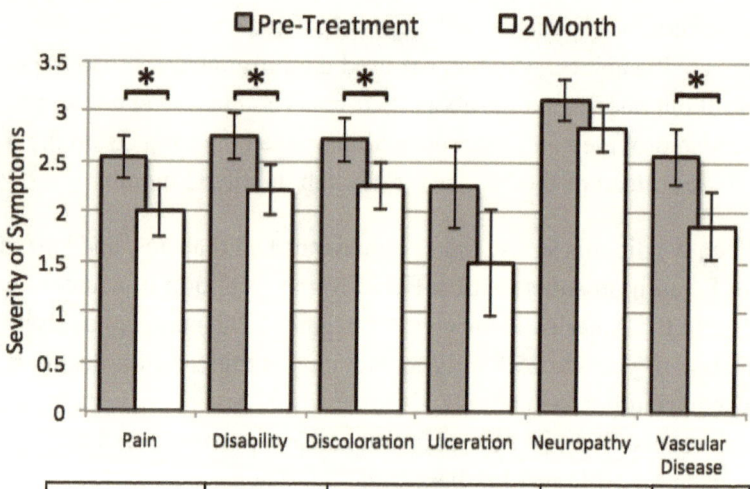

Symptoms	Number of patients with symptoms	Severity of Symptoms		Percentage Reduction	T-test
		Pre-Treatment	2 Month		
Pain	17	2.53	2	22.1%	**0.014**
Disability	16	2.75	2.22	21.9%	**0.0018**
Discoloration	16	2.72	2.26	15.1%	**0.0153**
Ulceration	8	2.25	1.5	41.7%	0.0796
Neuropathy	16	3.12	2.84	6.3%	0.2083
Vascular Disease	15	2.56	1.87	25.0%	**0.0217**

Changing the Protocol

At the end of two months, we reviewed the results. During the first two months, in addition to treating the bottoms of the feet, which we felt would engage the nervous system, the protocol involved treating both the pancreas and the thymus. The pancreas was treated at the front of the abdomen with the intention to treat the endocrine system at a point closest to the entry of excess sugars and carbohydrates into the body, presumably at the center of diabetic endocrine dysfunction. The circulatory system was treated by applying the Infratonic to the thymus, over the upper sternum. The intention here was to treat the circulatory system at a point where all the blood passes through the body every few minutes, carrying the value of stochastic resonance throughout the venous system.

We found that, while this initial protocol appeared effective in treating pain and vascular disease, it was less effective in neuropathy or numbness. To provide a better result, we chose to add a key point in the nervous system, the back of the brainstem, to further stimulate the nerves in the legs, from the brain down. Thus, at two months, the protocol was augmented to include applying the Infratonic to the occipital area of the head. This appeared to increase the relative effectiveness of the protocol with the symptoms of disability and neuropathy.

There was a complexity to this study which turned out to be quite valuable. At the time of the two-month study results, 3 of the 20 participants had dropped out, leaving 17, sufficient for good statistical significance. However, at the end of six months, another three had dropped out, or could not be reached. This left just 14 participants. In addition, five reported that they had discontinued the therapy. Of the 9 who continued, here are the results:

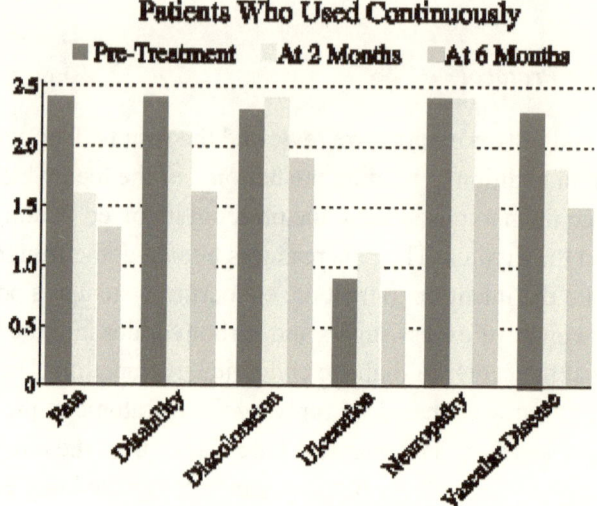

Patients Who Used Continuously

As you can see, all measures continued to improve, and the measure of neuropathy did improve in this group as anticipated from adding the brainstem treatment point. This was not the case with the five participants who discontinued treatment, as you can see below:

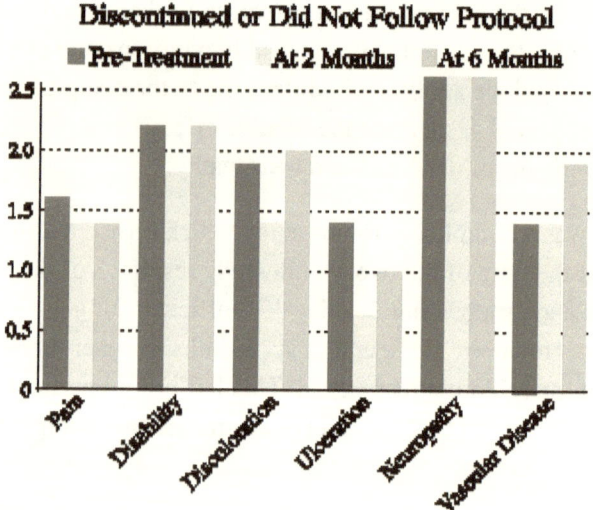

Discontinued or Did Not Follow Protocol

Those who stopped the treatment at two months also stopped, or began to reverse their progress in all measures! This clearly indicates that compliance is a critical issue in treatment of diabetic neuropathy. I have pondered this for a long time, and came to the realization that a big part of the problem was the need to treat each foot for 10 minutes with the Infratonic, and also to treat other points. This took almost an hour, and required bending over and repeatedly positioning the Infratonic on the floor with one or the other foot on top of it. If we were to have compliance, we would need a better way.

The Infratonic Footrest

The Infratonic footrest was born at the insistence of Suzanne, a meditation teacher of mine who believed that the Infratonic footrest and an Infratonic chair would be of great benefit. I was reluctant, but on reviewing this data from the neuropathy study, I saw some very clear benefits. With the footrest, the neuropathy treatment protocol is streamlined. With both feet treated simultaneously, they can be treated for 20 minutes, twice as long as before, while the other points are treated sequentially. Instead of a 40- to 60-minute protocol, this becomes a 20-minute protocol which includes twice the number of minutes on each foot. On writing of this book, the Infratonic footrest is in development. We have built and tested prototypes and have found that they are convenient to use and very effective at deep relaxation.

As I pondered why the Infratonic Footrest was so effective, I came across a published, peer-reviewed study in which electrodes were attached to the hands and feet, then connected electrically through insulated wires to a copper grounding post pounded into the ground outside. The results showed that, after two hours of sitting with these electrodes attached to their hands and feet, these test subjects showed reduced blood clumping, which means that the blood was able to flow more easily through veins and capillaries. They also showed increased zeta potential, which is a measure of the charge potential between the inside and the outside of blood cells. This, among other things, tends to cause blood cells to repel each other electrostatically, and not to clump. Now I had a second puzzle. How can connecting the palms and feet to earth through wires cause the blood cells to recharge in these ways?

There was yet a third puzzle. I have read studies which show that exposure to electromagnetic fields (EMF), such as exposure to cellphones, portable phones, computers, and low-frequency radiation from house wiring cause blood clumping, suggesting that blood clumping was a problem of stress induced by electromagnetic disturbance. How does electromagnetic pollution cause blood clumping?

I envisioned blood cells being beaten by the relentless EMF, then huddling together for safety. I envisioned those Qigong exercises that involve planting the feet firmly on the ground and connecting to earth ("grounding" in terms of most energy therapy methods) as extremely good at de-stressing and building up energy. I saw the act of connecting to the ground as creating a quantum mechanical wormhole from the chakras at the bottoms of the feet to earth. Such a wormhole would provide a conductive pathway by which this buildup of excess electromagnetic charge might then drain off into the earth. Wormholes are how I envision the functioning of my Scalene Light, to be presented in later chapters. Connecting the palms and feet to the ground electrically might provide a purely electrical alternative for draining off this excess charge to the ground and making the blood cells happy.

My objective is to provide the Infratonic footrest as a product that is ideally suited for diabetic neuropathy. A very important aspect of diabetes is high sugar level in the blood which causes a high level of acidity and inflammation. Thus, to fully address the condition of diabetes, we need to discuss diet and addiction to sugar.

The Biochemistry of Sugar/Carbohydrate Addiction

Like tobacco, alcohol and cocaine, sugar confuses the endocrine system into desperately wanting more by causing hormone levels to fluctuate wildly whenever the drug is stopped. Falling blood sugar level calls for another dose of the drug to bring it up again. Like caffeine and other addictive drugs, once you stop consuming it, your endocrine system normalizes and the cravings go away. After consuming high amounts of sugar and other carbohydrates, insulin level rise and, as a result, the glucose is pulled out of the bloodstream and stored as fat. When glucose goes low insulin levels drop.

Drops in glucose level cause desperate sugar cravings. As a result, the person consumes more and more, and the cycle continues. When excessive sugar and carbs are consumed over an extended period, and excessive stores of fat are already stored in the body, insulin stops going high so as to protect the body. Doctors say the blood sugar level is "uncontrolled" and prescribe insulin injections to get the body to store the excess sugars as fat. This reduces blood sugar level. However, it also stores the sugar as fat so the body will consume all the excess sugar and demand that we eat more. Low insulin and other hormones that go out of balance with it are the secondary cause of many of the symptoms of peripheral neuropathy.

There is a more detailed complication. High fructose corn syrup, now in most foods on the shelves of American supermarkets, contains glucose and fructose in about equal (and massive) quantity. Both go straight into the blood stream because, unlike sucrose from cane and beet sugar, a factory process has already broken up the carbohydrates into essentially pre-digested glucose and fructose. They just bypass the digestive system and are directly absorbed into the blood stream. This creates havoc because, while glucose is regulated with the insulin cycle, there is no mechanism for easy removal of fructose. So while the body might produce a high level of insulin to remove the glucose from the blood, high levels of fructose remain circulating in the blood. This makes the blood acidic and attacks tissues throughout the body, accounting for many of the degenerative characteristic of diabetes. Thus, high fructose corn syrup, which has come to dominate the processed food market over the last 20 years, correlates with the skyrocketing incidence of diabetes and neuropathic symptoms.

Refusing to produce insulin is just one of many ways the body tries to deal with sugar addiction. Insulin injections bypass this and cause excess sugars to be stored as more fat. Another way an intelligent system can sometimes remove excess carbs and sugars from the body is bulimia. Up to 1% of women in the U.S. suffer from bulimia. With forced vomiting after binge eating, people can eliminate much of the excess sugar and carbohydrate from the body before it gets into the blood stream. The body is very creative and attempts to compensate for sugar addiction. Our medical system refers to these attempts to restore balance as diseases, rather than taking the true illness head-on. It is sugar addiction.

One way sugar addiction calls for more sugar is through a perception of reduced energy level. "I feel tired. I need a chocolate or a protein bar, a latte, or something else loaded with carbs and sugar to get me going." Most people are entirely out of touch with the fat burning metabolism of the body, having fed sugar metabolism constantly, that their bodies simply don't know how to fuel the cells with fat. The fact is that people who regularly run out of sugar have long endurance and don't suffer from the sensations of low energy and food cravings because their bodies automatically shift over to the fat metabolism and provide fuel by drawing on the body's fat stores.

Western scientists believed, until relatively recently, that the body needed a constant supply of sugar, when a researcher traveling repeatedly to the North Pole noticed that their Eskimo guides didn't need sugar. They did great on a mostly fat diet. He studied this, and discovered that while it took an adjustment period, he could also eat a high fat diet with minimal sugar and carbohydrates, and have plenty of energy. This burning of fat for energy is called ketosis. Training the body is easy. It simply involves a dramatic reduction in sugar, and the willingness to suffer the inevitable withdrawal symptoms and often desperate sugar cravings. The benefits are huge. In addition to having increased vitality reserves and endurance, being able to burn fats for fuel means it is easy to maintain a healthy weight. It also offers rapid elimination of much inflammation, pain, and chronic disease.

Addiction is a very difficult habit to overcome. It takes a great deal of willpower to succeed. However, the recovery process is quick. 2 to 4 days of desperate sugar cravings, irritability, and headaches, and the worst is passed. Here are some effective methods once the willpower is in place.

1. The Apple Diet: Eat only apples for 3 to 5 days. Eat only apples. No applesauce or apple pie. Eat only raw fresh apples for 3 days. Ketosis will hit on the second day. The apple dieter will experience not only the above withdrawal symptoms, but also a bit of peer pressure. "Why aren't you eating with us?" Others will feel somewhat judged by the apple dieter. It is very easy for the pressure of addictive cravings to use the excuse of pleasing family or friends to break the diet. One baked potato, one order of fries, or a shake. One bagel on the second day will be enough to fill the body with a little sugar to defeat the effectiveness of the apple diet.

2. The Daily Fasting Method: Rats were fed a typical modern diet, and found to suffer from obesity, diabetes, and heart disease. Meanwhile, a second group of rats was fed the exact same diet, except that it was only available to them for 8 hours per day. Thus, every day for 16 hours, they fasted. Instead of being obese and ridden with illness, these rats were fit and healthy with no sign of heart disease or diabetes. The easiest way to implement such a diet is to eat raw vegetables every morning for breakfast with no sugar or carbohydrate. Then eat lunch and dinner. In this way, the fast is not broken until lunch, and the 8 hour dieter is fasting for 16 or more hours per day. Because the dieter is experiencing ketosis a little bit every day it can be more comfortable. Though it takes more sustained willpower.

    I usually eat a breakfast consisting of a small tomato, a few slices of cucumber, a little kale, beet, chard, basil, ginger, and lettuce, a small orange, a lemon, and a very small avocado. I add whatever vegetables I enjoy, sometimes cabbage, red bell pepper, or carrot. This provides a wide variety of vitamins and nutrients, living enzymes and vitality. I eat a light dinner and try not to eat after 7PM, thus am spending about 16 hours a day without sugar or significant carbohydrate. This 8 hour diet can be defeated by adding lots of calories to the breakfast or by eating a very early lunch or a very late dinner.

3. Another popular way is the "fast" which involves having nothing but water, water with a little lemon, only veggie juices, unsweetened fruit juices, or some other sort of very low calorie food for several days. Fasting can also be done one day a week, as in some religions, or continuously for 3 to 30 days. A minimum of three days is best because this is a typical amount of time needed to "break through" the addiction, completing most of the withdrawal symptoms.

After completing any of the above ketogenesis diets, most people find that they are no longer interested in sweets and their dietary life changes drastically.

The arteriosclerosis in blocked veins and arteries and the inflammation in legs associated with diabetes are mostly made of high energy fats which are some of the first to be eliminated during ketogenesis. This is why a month of gentle ketogenesis fasting (like the breakfast veggie diet above or not eating anything for one day per week) rapidly reverses heart disease, and diabetes, generally eliminating most inflamed fatty acid deposits throughout the circulatory system.

While I would like to sell lots of my Infratonic footrests to diabetes and heart disease patients for symptoms, the bottom line is sugar addiction. Without a determination to overcome this addiction, the Infratonic footrest might eliminate symptoms of neuropathy but the body will never be truly healthy until excess sugar and carbohydrate intake is curtailed. Here's a more detailed look at the mechanism of the ketogenesis process:

Ketogenesis

Ketogenesis directly and efficiently removes toxic, degenerative tissue. There may be no better, faster, or more efficient way to restore order to neuropathic degenerative tissue. Within a day of entering ketogenesis the acute symptoms of diabetes and neuropathy start to fade away. Enter the pathway of ketogenesis through regular fasting, and you have opened a doorway to abundance.

The two digestive processes for providing fuel to the cells of the body are normal digestion and ketogenesis. The first converts carbohydrates to release glucose. It is governed by the stomach and pancreas. The second process involves dissolving fat stores in the body through the liver to release ketones. Ketosis is a process in which the human body burns fats instead of sugars. Ketogenesis is the process of breaking the bonds between fat molecules in fat cells and having the liver convert them to ketones.

Ketogenesis is a rejuvenating process in that it scavenges stores of acidic, inflammatory fatty acids throughout the body, converting them to fuel. This cleansing does not happen during the more common sugar metabolism, so if people keep their digestive systems supplied with sugar and carbohydrates, these acidic waste materials never get removed. This causes chronic pain and inflammation, and many forms of chronic illness including diabetes, arteriosclerosis, and heart disease.

Ketogenesis scavenges the most accessible fats first. These are the high energy, acidic fats containing toxins, free radicals, and the byproducts of UV radiation damage, for they include substances which are less stable, less compatible with fat stores. When the call goes out for ketogenesis, these toxic and inflamed fats are the first sent to the liver for conversion into ketones, then sent back to the cells as ketones for burning as fuel.

This initial elimination of inflamed fatty acids is why one day of fasting is so effective at reducing pain, inflammation and other disease symptoms. This is an incredible rejuvenation technology which removes from the body most of the sources of chronic illness in the developed world. This is not new technology. It is this process of ketogenesis which reverses aging, and is why all the world's religions include, or at least included, a day or a week of fasting regularly. This is the science behind the spontaneous restoration of health and longevity induced by ritual fasting, and is why all the religions originally included fasting as a requirement for members of a spiritual community. Those who don't fast become a liability to the community. Those who do, find themselves in joyous service with vitality to spare. Ketogenesis is a new birth of freedom, rejuvenation, and abundance.

So, you may ask, "why does our modern world ignore the wisdom of fasting, and instead live by the myth of 'keeping up your sugar level'?" Why do we tend to snack on sugary things, and blame any sign of low energy on low sugar? If ketogenesis is so incredibly valuable to our health and longevity, why do most processed foods, most foods in a modern diet, contain lots of carbohydrates and sugars?

The answer is that drugs sell. Get people addicted to alcohol, caffeine, or tobacco and you have a captive market for your product. Add caffeine or alcohol to your product and you can get a share of the market. Most people in the developed world are addicted to sugar. This addiction is a very powerful one. Deny candy to a demanding child with low blood sugar and you can experience the wrath of an addict denied.

This should be astounding to modern medicine. However closed minded doctors don't notice dietary solutions to medical conditions. The open-minded doctors, and there are many, have known for years that diet can reverse heart disease, diabetes, and chronic pain. These progressive doctors have learned that it is most uncomfortable to come between an addict and his addiction. It is easiest just to mention something about eating a better diet with leafy greens, knowing it will be politely received, then ignored, overwhelmed by the addictive drive to consume more sugar whenever sugar levels in the blood start to dwindle.

Celebration

Dealing with excess dietary sugars is drudgery to the cells of the body. On the other hand, ketogenesis, the elimination of toxic and inflamed fatty acids, is a joyful, conscious celebration at the level of the cells. People doing this fasting become loving and joyful. Their bodies start to sing, that is, once they get through the withdrawal symptoms. Ketogenesis is a loving celebration as the cells have the opportunity to prune and nurture their stores of fat. It is also a celebration of accomplishment each time the person, as a whole, chooses this loving pathway, each time he feels hunger and chooses rejuvenation and revitalization over reaching for a power bar or a latte.

And bliss. As the stores of fat are purified and rancid, toxic, and inflamed molecular bonds are removed from the cell and sent into the bloodstream to be burned, the physical body purification is an enlivening increase in coherence of the cell membranes. This is experienced as bliss. This is a huge enlivening mission, to reclaim the body from illness. Every cell begins to sing.

An Intuitively Sourced Hypothesis:

It is easy to see how reducing consumption of sugars and carbohydrates eliminates symptoms of diabetes. However, the big question here is, how does the Infratonic, applied according to the neuropathy protocol, bring the endocrine system into balance despite continued availability of high fructose corn syrup and other sugars? I asked for intuitive brainstorming on this one. Here's what I got:

What does the Infratonic do when applied to the thymus and pancreas, along with the feet? *It engages the innate wisdom of the higher heart through the circulatory system to download a more effective operating system into the pancreas. This will be a system uniquely effective for the individual based on the experience of all of humanity.* [This mechanism is a bit of a stretch for most scientists but I like to pursue every lead, especially the more untested and exotic ones.]

Principal shifts to be expected from application of the protocol are generally the growing distaste for high fructose corn syrup and other sweets, a growing desire for exercise, and some method for cleansing excess stores of fructose and glucose from the body such as drinking lots of water. This means that the insulin and other sugar regulating hormones come into order and increase the previously repressed peripheral blood circulation which was responsible for most of the problems of peripheral neuropathy. [Many regular Infratonic users, including me, have developed a distaste for desserts, sweets, soft drinks, and anything with sugar in it. Thus, perhaps this explanation is really true.]

A suggestion to the user of the equipment and in some cases the test subject, will accelerate the process. Something like, "Over the course of the treatment you may notice changes like elevated mood and reduced appetite for sweetened foods. You may discover that foods such as those containing glucose and fructose may cause temporary cravings for more. Be aware of any such insights you get."

The Infratonic therapy will bring the wisdom of the endocrine system to physical consciousness without such a suggestion, but the process of recovery will proceed much faster and with greater consistency when the intuitive insights are supported with external verbal and/or written validation.

This gives us a novel research hypothesis: "Using the Infratonic Footrest with the diabetic neuropathy protocol will, by itself, eliminate or substantially reduce, sugar addiction." A second hypothesis, "A suggestion of awareness of sugar cravings during use of the Infratonic will further improve the outcome," can be tested once the initial hypothesis is proven. As always, "more research is necessary." It is one of the avenues I am pursuing because diabetes and obesity are such huge health problems.

II. Exploring Etheric Sludge

6. Chronic Illness is a Limiting Belief, an Unconscious Choice

Dissolving Limitation Leaves Us Open to Abundance

Abundance is our birthright. The only thing holding us back from abundance is limitation. What is this limitation? The Chinese refer to it as bad Qi, stagnant Qi, or pernicious Qi. They soften, dissolve, and remove these limitations to strengthen and normalize the flow of Qi in the body. I had studied this bad Qi and processes to remove it for many years, but it remained elusive, obscured perhaps in the Chinese language or Chinese medical mysticism.

Like a good scientist, I had broken Qi into 3 vital substances, the Chinese substances, Jing, Qi, and Shen. Each, I related to a different level or stratum of the ether, the electrical, magnetic, and gravitational, or from the standpoint of consciousness, the physical, emotional, and mental. This seemed somewhat helpful, but still had offered me little insight into just what "bad Qi" was made of. Then, I stumbled on the obvious. Bad Qi is information which disrupts the smooth flow of vitality.

We can view the Jing, Qi, and Shen to be 3 substances, like earth, water, and air. Water permeates earth, and air permeates water. This seems parallel to the fact that magnetic fields permeate electrical structures like atoms and molecules, and gravity permeates both magnetic fields in these electrical structures. Until the 1900's, it was believed that the whole universe, even empty space was filled with a substance which was called ether, within which planets rotated and light propagated. However, Albert Einstein, in 1905, discovered in his Theory of Relativity, that he could explain the functioning of the cosmos with or without ether. He believed in a principle which said that the simplest explanation is true. Thus he concluded that the ether doesn't exist. However, some form of ether is necessary to explain all this Qi and bad Qi phenomena. Whether bad Jing, bad Qi, or bad Shen, all can be called Etheric Sludge, residual, obstructing, limiting turbulences in the energy fields of the body.

This means that Etheric Sludge can be of different sorts, some of which cause congestion in the electrical substance, others, in the magnetic substance, and yet others, in the mental, or gravitational substance. In this section we will explore the details of each of these forms of Etheric Sludge. However, before going into detail it will help if we first look at an example of how it influences people's lives on a daily basis.

We are Programmed with Limitation

Every day, people engage in lifestyle habits which are damaging to their health. And yet they continue to do these things because that's what they are in the habit of doing. You're probably thinking of processed foods, sodas, drugs, and lack of exercise. This chapter examines less known, but more fundamental health damaging behaviors.

People frequently engage in mental chatter; self-judgment, questioning their own motives, doubting self-worth, expecting negative outcomes, etc. TV programming, the news media, and stories of disasters from friends and neighbors all contribute to the content of this mental chatter, and increase the expectation of pain, disease, and disaster. We are frequently told by our friends and family we are doing the right thing by watching the news, sharing disaster stories, and learning about chronic illness. "Did you hear about the 7 children who were murdered?" "We need to get rid of those terrorists." "Did you see the news this morning?" This reinforces the habits and the chatter. It also strengthens the disaster recordings in our heads and in our bodies that fill us with fear, anxiety, and depression, and cause the cells of our bodies to believe that we are likely to be victims of chronic pain and degenerative disease.

This phenomenon of dwelling on the negative is central to most of the chronic pain, suffering, and medical costs in our ballooning healthcare system. However, it is so endemic to our society that it goes ignored. How important is this programming to our health and well-being? This question is in desperate need of an answer. In fact, this field is being studied every day in great detail by capable scientists, but since the question is wrong, the answer is gibberish. The question is, "Is this scientific study influenced by the Placebo Effect?"

Studying the "placebo effect" simply shows us that giving placebo pills to patients and having doctors prescribe them is more effective than the rest of the medical system put together. In fact, most drug trials show that the placebo is far more effective than the drug at relieving the symptom. The placebo effect is responsible for 2/3 or more of the benefit. In fact commonly prescribed therapies, whether surgery or drugs, often prove no more effective than a placebo!!!! This is huge. I want researchers to get to the bottom of this, to get to the cause of chronic pain and illness that the placebo effect handles so nicely.

One huge belief, repeated over hundreds if not thousands of years is: "This is a world of limitation, of separation, illness, disease, doctors, hospitals, aging. We need doctors and drugs to compensate for the inevitable chronic illnesses." It's not like individuals have created all these ridiculous belief patterns on their own. We were born into a world where limitation, separation, and suffering are supported everywhere. We have been indoctrinated into the conscious and unconscious beliefs of the industrialized world. People connect to this limitation and suffering rather than connecting with abundance.

Chronic illness is just one small corner of the experience of limitation and suffering. People also manage to generate acute and chronic conflict with other people and repeat negative events along the lines of Murphy's Law, "Anything that can go wrong, will go wrong". Given that there is so much expectation of misery and suffering, it seems quite amazing that deep down inside, people feel love and joy. They tend to deny it, instead sharing stories of their own misery. Or, if they don't have fresh miserable experiences, they will share their own past misery, the misery others just told them about, or the misery they saw on television or read about in the paper. Misery, suffering, and limitation are a story that gets told so often in this world. And it is what is expected, not by all people, but by sufficient numbers to perpetuate most people's experience of it.

What would happen if people stopped *playing* the misery recordings and instead, *observed* the same recordings for what they truly are? An illusion! They would say to themselves, "My God! Look at the messes I have been programming into my life." If they could stay awake and really follow through on what they learn in this introspective process, they would be compelled to unplug from the negative feedback loops. They would stop watching, reading, and repeating the "news," stop watching TV commercials, dissociate from friends and relatives who parrot disaster stories, or more challenging, tell these "friends" that they don't want to hear the negative stories anymore. This will "destroy" many close and not so close relationships they have thought of as "friends," "family," and "loved ones." It will cause these "friends" to spend more time with others who are more willing to listen to these negative recordings.

A good question is: To what extent is our definition of "close relationship" based on sharing stories of separation, opposition, suffering, and illness? Also, how heavily are we criticized for refusing to listen, for refusing to participate in these negative conversations?

It is fascinating how offended people get if we refuse to allow them to share their disaster stories with us. Telling them we "don't want to hear it" is like telling them they have bad breath or they just told a lie. It is simply in bad taste to refuse to listen to someone's ongoing misery monologue. And yet, our bodies are ours alone, not theirs. We are 100% responsible for the programming our bodies, minds, and spirits get exposed to. Others don't need to live with the automatic misery, limitation, separation, and illness programs that play in our bodies. We do. So, what right do they have to feed their contagious misery programming into our bodies?

What right do we have to allow them, or to allow TV or the news to pollute our body, mind, and spirit? The answer is that *they* have no right, and *we* have 100% of the right to choose how our bodies will be programmed. If we choose not to choose, but rather to allow ourselves to be passive victims while all these misery, suffering, and illness recordings are poured into us, we are simply expressing our free will and choosing to be victims. This becomes our fulfillment of our preconceived belief that we deserve to suffer, that this is a world of suffering, and that suffering is inevitable.

There are endless examples of suffering in the world and it is easy to blame others for our suffering and disease, but it is all 100% of our choosing. The life we experience is 100% our creation, dependent fully on how we allow our bodies to be programmed. It might appear that lots of people are victims of circumstance. It can even be proven through scientific studies. For instance it has probably been shown that those who watch fast food commercials become obese, that those who watch doctor, disaster, emergency, and disease TV shows tend to get sick more often, and that those exposed to violence or crime in childhood tend to engage in the same experiences as adults. We can see this as proof that we are victims of circumstance. In so doing we are choosing to be that victim.

What is objectively true is not necessarily subjectively true. "They" may be governed by statistics, but I am governed by choice. Every day, every minute, I have the opportunity to choose my future by choosing what I am exposed to, and thus, programmed with. Others may make the unconscious choice to be victims, but because I have free will, free choice, and am creating my future every minute, I am the creator of my life, often choosing to pretend to be the victim, the sufferer, the chronically ill, alone and depressed. However, it is my free choice that has allowed this to come to pass.

Placebo Effect Revealed

As we realize that we are choosing between joy and misery every minute, we begin to see how we have created our present by pretending to be a hapless victim in the past. We also see that what we choose to expose ourselves to every minute is our method for programming whatever we choose into our lives.

So, psychosomatic illness, and really all chronic illness stems from choosing to be a victim of our lives and our environment, which fills us daily with the vibration and expectation of disease, drugs, and suffering. It fills us with a desire for those foods, lifestyles and information sources, which create chronic diseases. The placebo effect is a way to temporarily suspend a little of that programming, a way that feeds into the profitability of doctors, drug companies, and a myriad of alternative medicine approaches. However, it is temporary. Until we pull our power back from being a victim we will create the suffering again and again.

As we stand up and take responsibility for our own choices regarding programming our own bodies, we will find that our exercise activities increase. Our diet changes radically. Soft drinks and birthday cake become intolerably sweet. We surround ourselves with activities and people who uplift us. We will be more inspired, and motivated for a healthy, joyful, loving lifestyle. We will lose depressing old friends and gain delightful new ones.

Psychosomatic illness or chronic illness is a choice. It is not a choice made in the moment, but rather the product of years, or even centuries, of choosing to be the victim. However, the placebo effect reveals that spontaneous remission from chronic illness can happen at any moment, simply by suspending old programs, old beliefs in disease, and replacing them with expectations of spontaneous immediate relief and abundant health.

The existence of psychosomatic illness shows that, through expectation (usually years of programming), we can create a vast array of biochemical abnormalities in our bodies that are the physical manifestation of chronic illness, and that a shift in expectation through the placebo effect or other reprogramming methods can normalize ALL these biochemical indicators of disease. Spontaneous relief or spontaneous remission can result. With this awareness, it becomes absurd from a social health standpoint for pharmaceutical manufacturers to be attempting to balance the biochemistry of the body to "fix" chronic illness. From a corporate standpoint this is, of course, highly profitable.

You ask how researchers of the placebo effect and of nonspecific pain syndromes can research these effects. The answer is simple. By acknowledging that study participants have the free will to choose to create health or illness in their bodies, researchers can come to see how their test subjects, their experiment participants, have prepared the terrain of their programming for participation in the study. Test subjects (and many readers of this book) are massive storage devices of old baggage, who are expressing the illusions they have allowed their bodies to be filled with, along with the dietary and lifestyle choices which fulfill those illusions.

As an example, researchers might correlate subjects' disease conditions and past exposure to belief systems in sufficient detail to discover that their current disease manifestation stems from lifestyle choices, and that these lifestyle choices stem from the recordings or beliefs that they have been filled with. The correlation between disease conditions and lifestyle choices has mostly already been done. However the relationship between lifestyle choices, past experiences, and beliefs is mostly virgin territory.

Okay. What therapeutic methods are researchers most likely to find productive to study? Mindfulness is first to come to mind. It is also the most heavily studied in integrative medicine research circles. Learning to observe one's own life, the choices one makes, and the natural outcomes of making every choice consciously is huge in its potential to transform the medical system into a truly healthy health care system. It is also a productive way to open to abundance.

Etheric Sludge science is a very useful intermediate science for the present medical paradigm because jumping to a model of programmed beliefs being the cause of chronic illness is too big a gulf to cross for most traditional researchers. By studying the scientifically quantifiable substances of this negative programming that causes chronic illness, innovative researchers can establish a mechanism between mindfulness, conscious choosing, and relief from chronic suffering and illness.

Beliefs and habits have a measurable reality in the body, and can be erased and dissolved, not just with placebo filled talk therapy, mindfulness, and therapeutic prescriptions, but with electromechanical equipment which can be placebo controlled. Illness and suffering which are spontaneously relieved by placebo controlled procedures provide an intermediary link that is so needed by mechanistic doctors and researchers who are unwilling to jump the chasm between mind and body in chronic illness without repeatable measures of mechanism.

When it is shown that Etheric Sludge is a real substance that can be quantified by many modern techniques, and can be dissolved with other than "placebo containing" treatments, the loop is complete. It then becomes clear to all, and easy to teach, that Etheric Sludge is real stuff you store in your body when you are exposed to negative information, which, in turn, causes chronic disease and suffering.

For many, Etheric Sludge science will not seem palatable at first, just as "mindfulness" is held at the fringe of medical science. However as blinded, controlled studies prove its reality and explore its mechanisms, it will be accepted. As it is, chronic illness will begin to fade out of the mass human condition.

7. The Placebo Effect, Non-Specific Pain, and Chronic Illness

Research into the nature of the placebo effect has shown that positive expectations regarding a placebo therapy have, over the course of many experiments, caused virtually all biochemical markers related to chronic illness to normalize under controlled laboratory conditions. Even cancer and other "incurable" diseases have sometimes disappeared under the influence of the placebo effect. Further, about two thirds of the benefit of drugs, surgery, and other therapies appears to be due to the placebo effect. This is most remarkable. We could attribute this to the persuasive presentation by the doctor, or the patient's expectation when taking a magic pill. However, probing deeper, we start searching for what it is in the person's body, energy field or consciousness that can change upon the simple application of a placebo. A most interesting answer emerges. Just as the patient's beliefs and expectations can abruptly suspend nonspecific pain and chronic illness, it is the patient's beliefs and expectations that are creating the pain and holding the chronic illness in place.

Placebo Effect: the Elephant in the Room

When we look at most drug studies, we repeatedly see that the placebo effect is far more powerful than the drug effect itself, a fact often included in small print in the box with the drugs prescribed to us. The placebo effect often provides 2 to 3 times the relief or benefit as compared to the active ingredient being tested. Why do all these researchers continue to study all of these drugs when the placebo effect stands like an elephant in the room as so much more effective? Why do people focus on the drugs instead of this apparently marvelous, spectacular placebo effect?

This appears to be a huge important doorway into healing, except that "the placebo effect," points only to the idea that you will receive great health benefits if you take pills prescribed by experts. This is a most strange phenomenon. Why would the prescription of experts cause disease symptoms to disappear?

Focusing on the pill and the expert is like focusing on where the hands or the eyes of the magician divert our attention. It blocks a broader vision that would allow the opportunity to see what forces are really acting, forces that could possibly bring about such illnesses and discomforts that would be responsive to a placebo effect.

The Etheric Template

Deeper insight into this powerful phenomenon is achieved as we examine limiting beliefs from the standpoint of the etheric template. The etheric template is a pattern or morphogenetic field within and around the body, which causes all of our cells to re-grow perfectly just where they are needed. It comes into the body upon fertilization of the egg that becomes our bodies, and is a product of the Earth's evolution. It causes a single fertilized cell to differentiate and multiply into trillions of precisely positioned and functioning cells, maintaining them and replacing them as required. The etheric template can be seen in the following Kirlian photograph of a leaf after part of it has been cut away.

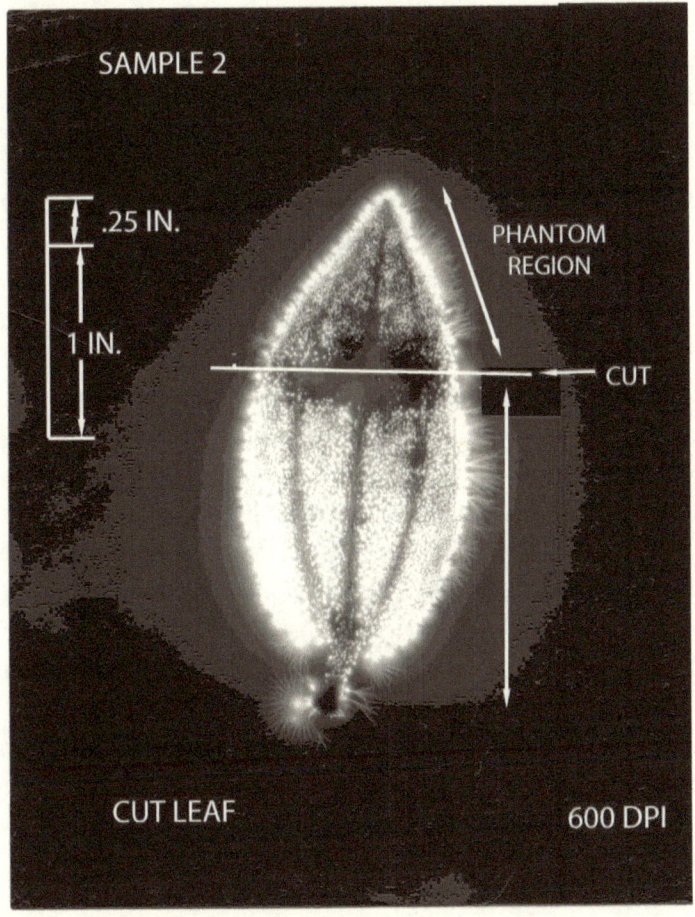

SAMPLE 2

.25 IN.

1 IN.

PHANTOM
REGION

CUT

CUT LEAF

600 DPI

   This photograph was taken by John Hubacher of Pantheon
Research, the first American researcher to reproduce the Phantom Leaf
results that came out of Russia in the 1960's and 70's. A leaf is selected,
then cut with a pair of scissors, and within a matter of seconds, placed on
a Kirlian Camera and an image taken. (Kirlian photography will be
discussed more thoroughly in a later chapter). The important point here
is that the etheric template of the leaf remains for several seconds after
the leaf is cut away. If it were a small injury it might heal completely.
However, a major cut like this one does not allow the etheric template of
the leaf to remain for long. In some cases of phantom limb pain in
returning veterans, the pain in a foot, is often felt for years after the leg is
amputated. Both the phantom leaf photographs and the existence of
phantom limb pain provide strong evidence for the existence of this

etheric template. As long as the etheric template functions perfectly we have no impeded healing and no chronic illness. This is the nature of the etheric template. However as will come as no surprise to most readers, it doesn't always function perfectly:

Our emotional and mental bodies are not the same as the etheric template. They are made of the same magnetic and gravitational substances, and exist in the same space as the etheric template, but they contain our thoughts and emotions. Depending on our thoughts and emotions, these two fields often don't cause the body to heal, but rather, confuse it, causing abnormalities. These two additional energetic bodies contain thoughts and emotions that can get stuck in the etheric template, warping it and causing it to create patterns of chronic illness and discomfort, rather than patterns of abundant health. A good example of this is a whiplash injury. Some people get into a car accident, suffer a whiplash, and recover quickly. Others get into a similar whiplash injury and don't recover for months or years. An important difference is what emotions the person was feeling when they experienced the accident. Whether the thought was "it's my fault" or "the other person is to blame," or "now my life is ruined," whether it is self-judgment or rage, emotion gets confused with the etheric template in the area of the injured cells. Instead of proceeding toward rapid healing, inflammation, congestion, or muscle spasm often emerges and doesn't go away. This is a case of emotion or judgment getting enmeshed in the etheric template.

The etheric template is the fabric of our amazing healing abilities. It is the source of all healing, of spontaneous remission. It also makes the placebo effect possible, but only when limiting beliefs and dysfunctional emotions become woven into the etheric template. Thus, just as the etheric template can cause the body to proceed toward rapid healing, when it gets intertwined with dysfunctional thoughts and emotions it can cause the body to remain ill, and when the placebo effect, the "power of suggestion," suggests that we should immediately recover, it's simply a case of temporarily withdrawing the foreign material from the etheric template to induce a spontaneous remission from symptoms.

Limiting Beliefs

What are these limiting beliefs that get stored in our etheric template that might cause chronic pain and illness, and might disappear or go into hiding when a doctor prescribes a pill? Here are some possibilities:

I don't deserve...
I deserve to be punished...
I deserve to be in pain...
I am not worthy...
I can only experience healing if an expert grants me permission...
I can avoid taking responsibility for my life by doing what others tell me to do. Then if it doesn't work, I can blame them for my suffering...
I can control others by suffering when I follow their suggestions...
I make mistakes...
I never get it right...
Suffering and illness keep others close to me...

For many generations the masses of humanity have been controlled by these limiting beliefs that reside within them. We believe we cannot be healthy, correct, or successful unless we rely on the expertise and permission of those who we perceive as superior to us. Essentially, we believe we cannot heal our own bodies, cannot function correctly, unless we pay experts in exchange for their expertise and permission to heal. This huge body of limiting beliefs is so embedded in our etheric template that it causes our illnesses, health limitations, and performance limitations. Further, it causes premature aging because we believe aging is natural, and expect it.

Thus the placebo effect, which focuses our attention on the doctor in the white coat and the magic pill, tends to block us from the realization that limiting beliefs cause our chronic illness and suffering in the first place, and focuses our attention solely on the belief that non-specific pain and chronic illness are normal and that we need to have experts prescribe pills for us to keep these chronic conditions under control. This is a huge limiting belief.

119

As we find these limiting beliefs within us and manage to pull back our power from them, we eliminate them from our etheric template and they lose their power over us. The idea that "I need doctors to heal me," is very old and deeply embedded. It allows us, through the placebo effect, to temporarily override our negative beliefs and subconscious limitations. However, it does not eliminate them. Instead it reinforces the belief that the doctor is necessary. It permanently installs the doctor or other professional more deeply into our etheric template so that we depend upon them instead of ourselves for our healing and well-being. The more we rely on a doctor and a pill, the stronger the limiting beliefs become, and the more likely we are to experience unexplained chronic pain and need the doctor and the magic pill in the future.

Holding these limiting beliefs in our etheric template only benefits others at our own expense. Releasing these limiting beliefs opens the door to spontaneous healing, spontaneous remission and the disappearance of chronic or life-threatening conditions. Thus, we can see the difference between the placebo effect and the removal of limiting beliefs from our etheric template. The former, disempowers and the latter, empowers. The placebo effect provides temporary relief, while the process of releasing limiting beliefs offers long term, if not permanent, relief.

Thus the placebo effect, this apparently spectacular elephant in the room, blinds us to the true cause of our suffering, and puts the doctor in control of our lives. Our accepting of the placebo effect assures that we will manifest an endless chain of diseases which can be treated by the placebo effect. However, clearing limiting beliefs from the etheric template provides us freedom from the expertise of others for our health, healing, and vitality. You will see in a later chapter that the centenarians of Vilcabamba, Ecuador lived to be 120 and 130 years old, healthy and happy to be working in the fields well past 100 years of age without access to anything remotely resembling our modern medical system. Thus the placebo effect, which appears to be such a marvelous discovery, is really just one more limiting belief which keeps the doctor in control and assures that we have little power to heal ourselves.

Have you ever had a disease or discomfort and headed to the doctor to get a prescription, then noticed that your symptoms were already disappearing even before you take the prescribed drug? This is a common experience. It is clear from this that you had given away your power, and your symptoms were created by the limiting beliefs embedded in your etheric template, awaiting a doctor's permission to disappear. In truth you had the power to heal yourself all along. This is the real mind-blowing elephant in the room.

Proposal for Placebo Effect Researchers

The placebo effect has long been considered by researchers as a confounding variable, associated with pills and doctors' prescriptions. We believe this has blinded many researchers to the true importance of the placebo effect. The placebo effect appears to be based on limiting beliefs impressed on the etheric templates of people by effects best understood as Pavlovian training. The following are two proposed research hypotheses:

Subconscious limiting beliefs are the problem for which the placebo effect provides temporary relief, and

The placebo effect can be induced using Pavlovian conditioning.

As evidence that the placebo effect has an important component that lies outside of our thoughts and beliefs and resides in the physical body, the following two studies show that placebo effects can be embedded into rats and mice through Pavlovian means of positive and negative reinforcement, and become automatic responses of the body.

A study by R. J. Hernstein published in Science entitled "Placebo Effect in Rats" reports: "Scopolamine hydrobromide disrupts the learned behavior of rats in a predictable manner." Physiological saline mimics to some extent the effect of the drug when the two substances are alternately administered in a series of injections. This placebo effect appears to be an instance of simple Pavlovian programming."

A second study, "Dissection of Placebo Analgesia in Mice" by J.Y. Guo, published in the Journal of Psychopharmacology" goes further. It reports; "Placebo analgesic responses were evoked by an exposure to a conditioned cue previously paired with drug conditioning. Morphine conditioning produced placebo responses that were completely antagonized by naloxone. By contrast, the conditioned cue after aspirin conditioning elicited a placebo effect that was not blocked by naloxone." In other words, whether a particular drug works to relieve a symptom depends, not on the symptom, but rather on how the Pavlovian programming was induced. This shows us that drug responses are easily shaped by Pavlovian programming.

Have you ever seen a TV commercial in which a suffering person is shown as totally miserable with symptoms of a common cold or allergy or other painful or embarrassing symptoms? Modern day TV ads to sell drugs provide emotionally evoking visuals that create a feeling of suffering while describing the symptom, and asking "Do you have this condition?" The implication is that this condition is common and that the audience is likely to suffer from it. This is followed with "Ask your doctor if this drug will relieve your symptoms." This is pure Pavlovian programming: evoking an uncomfortable emotion, then suggesting that the audience has the symptom, evoking the suffering some more, then telling the audience that a new drug, prescribed by a doctor, will relieve the symptom. *Done well, these commercials (which used to be in violation of law) can implant both the disease and the placebo effectiveness of the drug into millions of people, all through Pavlovian programming.* As the above animal studies show, even a pure placebo, presented in this way, will be effective at relieving symptoms that were caused by Pavlovian programming in the first place.

University of Connecticut psychologist Irving Kirsch studied the effectiveness of antidepressant drugs with a meta-analysis of 19 antidepressants and found that 75% of their effectiveness was due to the placebo effect. His results were published in "Prevention and Treatment," a peer reviewed journal of the American Psychological Association. He went further, arguing that even in the remaining 25%, the "real" drug effect appeared to be a disguised placebo effect because the "real" drug has obvious side effects like dry mouth, nausea, dizziness, or sexual dysfunction, whereas the placebo does not. Some studies showed that up to 80% of patients could guess correctly to which group they were assigned. Clearly, these studies were not blinded. Through Pavlovian training, we have all learned that powerful drugs have powerful side effects. Thus, if we get side effects, we believe a drug is working. Since most of the subjects knew which group they were in from the side effects, this could easily account for the remaining 25% effectiveness of the drug over the placebo.

Supporting this finding, Roger Greenberg at the State University of New York Health Science Center in Syracuse did another meta-analysis study of psychiatric drug trials and found that the severity of the side effects was correlated with the drug's efficacy. Those test subjects who experienced more side effects got better therapeutic effects from the drug. Thus again, it appears that the effectiveness of psychiatric drugs may be almost entirely due to the placebo effect.

Rather than a valuable treatment modality, drugs, along with the Pavlovian programming that comes with their promotion and administration, appear to be causing the limiting beliefs that cause chronic illness and psychological disorders, paving the way for the placebo effect to relieve them.

The limiting beliefs that cause psychiatric disorders, chronic illness, and impeded healing can be induced by traumatic childhood experiences, parental or societal beliefs. They can also be induced by the media's obsession with illness, and by mass advertising. Doctors, therapists, nutritional supplement salespeople and network marketers all teach people that disease is normal, and often inevitable with aging. They also teach that their product is needed to overcome illness and symptoms from which their customers are likely to suffer. This instills fear and the expectation of disease in the public. It also instills the need for therapy, particularly pills, to prevent or recover from precisely those diseases they fear.

The dominance of the placebo effect is not limited to psychiatric disorders. The effectiveness of many surgeries and drugs that are part of mainstream medicine, when tested with blinded placebo-controlled protocols, have been shown to be due entirely to the placebo effect. In the book, THE DIVIDED MIND by John Sarno, a survey of meta-analyses showed that there was no solid support in the literature that any of the popular treatments for low back pain are effective. A meta-analysis is a paper which analyzes and discusses the results of many published papers. Whether exercise therapy, lumbar supports, TENS, or even decompression surgery, none were found to be generally effective. Remaining physically active was advised because of the harmful effects of excess bed rest. Also, psychological behavioral treatment was found to be effective. Despite these findings of non-effectiveness of physical medicine for low back pain in the literature, it remains that, in our society, cutting somebody open with a scalpel, injecting them with a needle, or consuming a drug that creates side effects is powerful medicine. This makes modern medicine a very powerful placebo.

Getting to the Bottom of the Placebo Effect

How might this Pavlovian programming aspect of the placebo effect be studied?

First, a researcher might correlate prevalence of chronic illness with exposure to media and confidence in doctors. Do people who watch TV and read articles about disease have more chronic illness than those who don't?

A survey can be conducted which asks participants a battery of questions to quantify not only the extent of their chronic illnesses, but also quantifies risk factors. Participants might be asked about hours a day spent watching TV and reading the paper, the extent to which medical TV programs are watched or medical articles are read in the media, or the trust that these participants have in medical doctors, alternative practitioners, and nutritional supplements. They might also be asked about their confidence in their own ability to remain healthy without pills. These are just some of the factors that might prove to be important risk factors in chronic illness, impeded healing, nonspecific pain and psychiatric disorders.

Research is needed to compare those subjects who respond to a particular placebo, with those who don't. What is their background? What are their belief systems? What factors cause these people to become ill with these particular symptoms such that they need therapy? And what causes this therapy to be effective with them? This is a very broad and important study because it is likely to show that limiting beliefs and Pavlovian programming are responsible for the majority of chronic illness today.

It may be difficult to identify limiting beliefs in psychological questionnaires because it appears that these limiting beliefs are subconsciously programmed by Pavlovian programming. Programmed subconscious beliefs may be very different from patient expectancy. On the other hand, a study of the outcomes that patients expect before a placebo effect study might offer an important window into limiting beliefs. One way to go about this would be to have patients complete a questionnaire designed to profile their limiting beliefs and their expectancy, then to administer a placebo which is claimed to relieve their symptoms, and finally to repeat the questionnaire to see the extent to which their limiting beliefs or their expectancy determines the outcome.

A more mechanistic study involves quantifying the ways in which these Pavlovian reactivities are stored in the body, such that they might be better understood and effectively removed. Here we hypothesize that Etheric Sludge is the substance of limiting beliefs, the vibrational signals that ride on the substances of the etheric body, which have been embedded by Pavlovian programming. Many therapeutic methods provide relief from symptoms that last a few hours, days, or weeks. Symptoms then return and the patient requires repeated therapy. Here are some popular methods for removing these limiting beliefs:

Tapping and EMDR

Emotional Freedom Technique (EFT), often referred to as "tapping", is one example of many therapies that are claimed to remove these programmed reactions from specific localized areas of the body. Eye Movement Desensitization Routine (EMDR), is another. Both depend on the theory that Pavlovian limiting beliefs are stored vibrationally in the physical body. Tapping applies vibration to specific points to help break loose these buried limiting beliefs. EMDR uses physical movement of the eyes to break up these beliefs. There is no question that they often get results in a clinical setting. However, the confusion between the placebo effect and practitioner administered therapies is so huge in these techniques that it is very difficult to study them with blinded controlled protocols. Thus they don't have the credibility for scientific investigation.

John Sarno, M.D., and Psychosomatic Disorders

Two of my brothers tell me that they used to have chronic pain. Then they read a book by John Sarno, M.D., and just from reading and understanding the concepts in the book, the chronic pain was gone. As I looked into Dr. Sarno's work I discovered that he believes that, to achieve long-lasting relief from pain and suffering, it is important to address the underlying limiting beliefs. Examples he offers are:

"People with my genetic makeup tend to get this disease."
"I cannot heal by myself."
"I need the intervention of a highly educated and trained expert."
"Aging is inevitable along with pain and chronic illness."
"These conditions are incurable without the help of a medical professional."

He argues that these beliefs, stored at a cellular level and not accessible to our conscious awareness, directly cause the pain and suffering. They are only relieved when someone intervenes with authority and tells us we can take a pill or get a procedure to get rid of it.

Unless these underlying limiting beliefs are dissolved, the seeds of chronic illness will remain, and either the current condition or another disease syndrome will return in the future.

His method, put in a nutshell, is simply to tell patients with chronic pain that their pain is psychosomatic, and to explain to them how their body is capable of producing chronic pain through belief and intention. He does this, of course, while wearing a white coat and having an M.D. after his name. Thus again, his success with my two brothers, and his success with his patients, while convincing, cannot be separated from the placebo effect. It is simply not scientifically persuasive to study techniques embodying the placebo effect to conclude anything about the nature of the placebo effect. Thus Dr. Sarno's work, and the work of a whole series of placebo effect doctors over the years, goes ignored by mainstream placebo researchers, or at least by the peer reviewed publications that control what papers are published.

Removing the Practitioner

Health practitioners can have a huge influence on outcomes regardless of whether they wear a white coat and have an M.D. after their names, simply by claiming authoritatively that the patient will or will not get better. In order to study the mechanisms of the Pavlovian placebo effect, it is generally necessary to remove all practitioners from the study. For this reason, we propose to utilize several electronic devices, each of which removes a particular kind of Etheric Sludge. All of these devices can have sham devices created that closely resemble them. One may prove more effective than others depending on the form of limiting belief or Etheric Sludge, the subject's background, the condition, or the particular effectiveness of the device. These devices and their principles of operation will be described in the following chapters.

Landmark Placebo Research

Here are two proposed hypotheses: "Unconscious limiting beliefs stored in the body are the source of the placebo effect. If the substance of these limiting beliefs is removed from the body the placebo effect will not work." A secondary hypothesis is: "If limiting beliefs are removed from the body through electro-mechanical means, symptoms of nonspecific pain and chronic illness will disappear spontaneously."

This can be studied with animals to test whether Pavlovian programming can be removed with electronic devices, and in human studies to quantify the relationship between Pavlovian programming and the placebo effect.

Animal Studies

The first kind of study proposed to test this hypothesis is done with animals wherein they are first conditioned through Pavlovian techniques to have a response to a placebo stimulus. Then one of the groups receives a therapy intended to remove limiting beliefs. The placebo stimulus is applied to the control group. The hypothesis is that the treated group will have less of a placebo response than the untreated group, indicating that the effect of the Pavlovian training has decreased. It is important to test soon after the administration of the limiting belief removal technique, then hours, and days later, to discover whether and how quickly and strongly the Pavlovian training returns to control their behavior again. Different techniques and devices can be employed for this test. The following chapters describe some proposed equipment.

Human Studies

For humans the strategy would be more complex. Subjects can be recruited from a group with a condition that responds often to placebos such as arthritis, back pain, or fibromyalgia. The subjects are paired and divided into 2 groups. A questionnaire is administered repeatedly during the study to determine the subjects' level of pain or discomfort and to quantify the subjects' limiting beliefs and the extent to which their lives are controlled by these limiting beliefs. The questionnaire is repeated to see how the strength of these limiting beliefs varies at each step of the test. The treatment designed to remove limiting beliefs is applied to the experimental group whereas a sham treatment is applied to the control group. They are tested again with the questionnaire to see whether their symptoms or their limiting beliefs have changed. Then a traditional therapy with high placebo effect is administered and the questionnaire is repeated a 3rd time. If the preconditioning is successful at reducing the unconscious limiting beliefs, the placebo will have less effect. At the same time, we may see spontaneous remission of health conditions, or initialization of recovery, caused by elimination of the limiting beliefs.

The following specific devices, which work in different etheric substances and on different forms of Etheric Sludge, will be described in the next chapters:

1.) Vital 1 to reduce anxiety, fear, and susceptibility to external fields.
2.) Scalene Light for depression, grief, worry, and other mental conditions.
3.) Infratonic S for dissolving limiting beliefs stored in the area of the upper abdomen, diaphragm, and chest.

The Infratonic Footrest, which reduces stress and low energy magnetic charges in the body, was described in a previous chapter.

A Landmark Study

This is a landmark study which, if effectively implemented and disseminated, will awaken the medical community to the true cause of the placebo effect, nonspecific pain, and chronic illness. It will show that the cause of most disease and suffering is energetic substances which have a physical/informational reality. The fact that these substance can be removed through electromechanical devices will show that it is not just "in the head" and that it is not to be found with microscopes in the physical structure of the body either. It is in the energetic fields that surround and interpenetrate the body. This study will show that Etheric Sludge is the cause of most human suffering, and that releasing the body from the limitations imposed by Etheric Sludge will open the door to abundant health.

8. The 3 Vital Fields with Devices that Illustrate Them

To understand the concept of Etheric Sludge, the reader needs a background into vital fields, or at least my version of vital fields, and how they fit together in an overall picture of human health, wellbeing, and consciousness. From here you will be able to understand why I developed the devices that I did, and how I believe they function.

The Electric Field

First there is an electric field that surrounds the body up to about an inch from the skin. Activity in this electric field can be measured in EEG, EMG, EKG, etc. However this electric field is more than just electrical impulses produced by organs and neurons. I used to travel extensively with Dr. Fu, who was highly sensitive to airplane travel. We were also leading tours of groups of doctors and therapists to China. This gave me a great opportunity to research the body's electrical field and how it is influenced by air travel. I identified those people on the airplane who were highly sensitive to air travel, and put different devices on them. I then measured their energy field with a Kirlian camera and asked them for their qualitative experience of the flight. I learned that magnets, applied to the body, are effective at reducing symptoms of anxiety, spontaneous sweating, headaches, and rising heat. I found that air ionization was effective at energizing the body to reduce symptoms of exhaustion and a depleted immune system that are characteristic of long airplane flights.

I put these together in combination with unpredictable signals to create the Vital 1, a tiny device that you can put in your pocket while traveling. It is very effective at increasing comfort during a flight and reducing or eliminating the symptoms of exhaustion and a weak immune system that often results in catching cold or flu after a long flight. This

Vital 1 has additional uses. Many energy sensitive people find that Walmart and other places with big crowds and fluorescent lights can make them very uncomfortable. They find that wearing the Vital 1 allows them to feel comfortable and able to shop without feelings of anxiety.

This led me to a theory of strengthening the etheric template. It appears that the Vital 1, by inducing unpredictable pulsations of an electromagnetic nature within the electric field of the body causes the etheric template to intensify and become organized such that electromagnetic influences and thought forms from people don't penetrate into the etheric template, and even drives them out. This concept or theory has many applications. One of them is the discovery that the Vital 1 appears to treat chronic cough effectively. I have found 6 people over the years that had a chronic weak cough for more than 2 weeks. I asked them to wear the Vital 1 for 48 hours continuously. In all cases the chronic cough was gone by the end of the 48 hours.

Not long after I developed this technology, I had a chronic cough for a month, and was really uncomfortable about getting on a plane for a long flight to Australia. I thought I was going to continue coughing throughout the trip, making everyone around me fearful that they would catch some awful disease. As it turns out, I put on the device when I got on the plane and it seemed to wake up my lungs. For the first ½ hour of the flight I was coughing excessively with a productive cough, definitely worse than I had feared. However, after that first ½ hour, the cough dissipated and did not return throughout my two week visit to Australia.

This might be explained in a variety of ways: One of them was my original theory: that the electrical and magnetic impulses in the electric field of the body increase the electrical and magnetic vitality of the body, which strengthens the immune system allowing it to fight off whatever bugs or viruses are present in the body. The "etheric template" model is similar except that the mechanism is a densification, an organization of the etheric template. As the etheric template becomes denser and more coherent, it drives out alien forces like electromagnetic and emotional influences, the effects of emotional crowds of people, anxiety, and diseases like viruses that are not a part of the etheric template. These influences tend to be pushed out as the etheric template gets stronger. Thus, this understanding of the Vital 1 would appear to open the door to a whole new paradigm of medical research, that of bringing healing to the body by strengthening the etheric template, and driving out external disease factors.

The Magnetic Field

The second field of the ether is the magnetic field, the emotional body. This is where the Infratonic devices, low frequency vibrational devices that apply massage action in the range of the brain's Alpha rhythm to ailing parts of the body, appear to be most effective. They calm and normalize excesses in the activity of this magnetic body. They also unravel blockages in this magnetic body, which releases energy that can catalyze healing. A big key to their effectiveness is that they produce unpredictable signals in the range of the nervous system, and thereby access and unravel distortions in this magnetic field. The magnetic field of the etheric body is measurable through magnetic sensors such as SQUID (Superconducting Quantum Interference Device) a very sensitive magnetometer used to measure extremely subtle *magnetic* fields. Even though this field is measurable, the idea that this field contains information of consciousness is at the forefront of science. The connection between this field and emotional information has not been accepted by most medical researchers. This leaves it as 'virgin territory' for intrepid scientists.

Another device, which works in this second, emotional field, is the Infratonic S. It produces coherent linear magnetic pulses that are more effective at engaging stuck emotions. Subconscious limiting beliefs are good examples of highly resilient distortions in this field. The etheric template is an ideal healthy pattern for the physical body. Our thoughts and emotions are not natural to this etheric template and distort it. Limiting beliefs are powerful influences within us that are not "true" from the standpoint of the etheric template. They create distortions to the etheric template. This technology has vast applications, including those for the military, both for unleashing optimum performance by dissolving limiting beliefs, and for deprogramming Post-Traumatic Stress Disorder (PTSD), which are highly traumatic experiences that get embedded in the etheric template.

The Mental Field

The third field is the mental field, something like the gravitational field, except that the gravitational field is perceived as only existing in straight lines radial to the center of the earth. This mental field might be viewed as slight turbulence in earth's gravitational field. Data stored in the gravitational field may allow for a huge amount of information, the substance of mind. Also just as gravity can bend the path of light and magnetic fields, this mental field, in this model, is able to shape the emotional field. The Scalene Light produces a Ray in the substance of this mental field, and modulates this Ray so that it interacts with and disperses what we might call thoughts in the field that surrounds people.

While the Scalene Light is effective in a wide variety of applications, it is specifically pointed out here because it is so good at dissolving the mental congestion that interferes with mental clarity and offers fast relief from symptoms of depression. In preparation for this book I spoke with three psychologists who have the Scalene Light and found that they use it to clear their counseling room after each client. The therapists felt much better, and they found that their next client could relax much more quickly, would get down to productive work more quickly and made more progress in these sessions. It is simple to set up a blinded controlled experiment in which depressed test subjects are tested with a questionnaire then asked to walk through one of two randomly selected corridors. Scalene Lights can be hidden along each corridor such that experimenters and subjects don't know which one is the placebo corridor and which is the treatment corridor. This one experiment, repeated extensively, will radically change how psychiatry and counseling are practiced.

The biggest value of the Scalene Light in my mind is that it teaches us that we are not our thoughts. It teaches us that we put our thoughts in place around us and our thoughts then echo our past state of mind back to us. These echoes are not us. They are simply residual thought forms that cause us to repeat and sustain thoughts and states of mind that we dwelled upon in the past. In addition, we discover that we are strongly influenced by the residual content of spaces that we occupy. This teaches us the importance of mental pollution, and mental hygiene.

Differences between Anxiety, Stress, and Depression

The definitions of these three words often vary and overlap in confusing ways. For the purpose of this book, I have defined them quite specifically to make clear distinctions between them. I have defined them to clarify the distinction between electrical, magnetic, and mental substances. From this perspective, anxiety is caused by subconscious fears and the fight or flight response. Stress is caused by frustrated wants and demands or pressures imposed by the environment. Depression is caused by dwelling on hopelessness, despair, and other repeating thoughts. We might include grief and other conditions of dwelling on thoughts, but for clarity, we'll stick with just depression.

The substance of anxiety, fear and survival programming from prehistoric times, is energetic programming stored in the electrical body, specifically, the sympathetic nervous system. The substance of stress is energetic free radicals, high energy particles in the substance of the emotional body. Depression is associated with a cloudy mental substance that interferes with nerve conduction at the synapses and encourages dwelling on the same thoughts over and over. Anxiety, stress, and depression can be relieved in controlled placebo studies by the Vital 1, Infratonic footrest, and Scalene Light respectively.

These are some pretty bold generalizations. I've tested them out, and have discovered, amazingly, that they seem to be true. People with anxiety in crowded environments can hold a Vital 1, and within seconds, feel comfortable and confident. Within 10 minutes of placing their feet on the Infratonic footrest, people discover that much of their stress is gone. Psychologists have used the Scalene Light in the room with depressed clients, and have found that an overlying layer of depression lifts within a minute or two. In all cases, how quickly it returns is determined by the environment and the thought processes of the individual.

Early Morning Writing

You may wonder where this idea of Stress, Anxiety and Depression being treatable by three different devices came from, and why I seem so confident about it before fully testing it out. The answer is that I wrote it down at 4:00 AM a few months ago, and when I read it later that morning, I realized it made a lot of sense. I proceeded to test it out to the best of my ability and thus far, it has proven true. You will read about how I started doing automatic writing later in this book, but, for now, I should just tell you that what follows, and much of the writing you will encounter through the rest of this book was written in the early morning, inspired by some higher consciousness. As you will see below, my writing goes much further, providing lots of testable details only a few of which I have had time to validate. *From here on, when I am presenting what I wrote, I will put my questions and commentary in italics. Here is what I wrote:*

4:00 AM

None of these treatments will be believed until validated. Even then, there will be great turmoil as people tend to continue to believe their stories, even when faced with alternate experience and evidence. This is particularly true where people have a vested interest, either in their disease or in their treatment method. In addition, people will resist these methods because they appear to be impersonal, devoid of patient validation, discussion, and analysis. The problem here is the mind or ego of the psychologist, lay counselor, or clergy. The ego tends to believe that all solutions in psychological areas stem from analytical or cathartic processes.

It is true, and shall be considered remarkable. Each of the conditions shall be easily treated with each of the devices, respectively. Depression and anxiety receive immediate benefit, whereas grief, rage, and betrayal require a longer treatment. Treatment of these conditions does best followed by the Infratonic. In fact, all do and people feel better when the therapy is followed by the Infratonic.

Anxiety returns in an hour. Depression returns in 2 to 3 days. And anxiety, grief, rage, and betrayal are lessened with each use. One, three, or five treatments are often effective. Yet, one treatment and discussing the object of rage is as effective as three treatments, using the Infratonic S alone.

Anxiety has a visible or hidden source of fear, which calls for continued vigilance. Emotions of grief, rage, and betrayal have been attached to the body, which has allowed people to become disconnected from God, a loved one, or an object like a car, home, or cell phone. Depression is simply separation from God, and the Scalene Light temporarily erases that separation such that Union is restored. This will be a tough sell because so many don't believe that such a relationship with and disconnection from God is even possible in this way.

1.) The emotional processes, grief, rage, and betrayal, respond well to catharsis, but this is not generally so for anxiety or depression.
2.) Fear is emotional.
3.) Anxiety is physical.
4.) Abandonment is a blend of grief, betrayal, and depression.

Hybrid issues like abandonment are excellent because pre-and post-questionnaires will clearly show the difference between the Scalene Light, Infratonic S, and Vital 1.

The best research would involve testing large numbers of people with a wide variety of conditions, applying one of the modalities at random with a blinded protocol, and then retesting. Results will show the specificity of the interventions. This is a much better test than pre-sorting people by condition because the act of pre-sorting will be subject to a great deal of scientific objection.

What is pain? You will find there is also physical, emotional, and mental pain, and people have a mixture of all three. The same device will work for each category, respectively.

Anxiety is caused by hyper vigilance of the sympathetic nervous system based on the fearful limiting belief that something bad is going to happen. This calls for cortisol which readies the nerves for action.

The Vital 1 awakens the parasympathetic function, and puts the sympathetic to sleep. How does it do this? This occurs by anticipating pulses in the 1 kHz to 3 kHz range. This is too fast for the sympathetic nervous system. The hyper-vigilance is focused on sensing the next pulse, and loses track of the next bad thing that is feared.

The parasympathetic is the system that perceives the pulses. Vigilance shifts to parasympathetic. Give someone a massage and they can't be vigilant because they are drawn into parasympathetic anticipation of the next touch or stroke. The sympathetic nervous system is separate from the cerebral spinal column. This is a huge clue as to why humans feel separate and lost. The ego is this collective cognitive awareness that averages everything together and maintains limiting beliefs.

A Cure for anxiety No coffee/caffeine or other stimulants; a veggie diet with lots of omega 3's; use of the Vital 1; and patience. When following the procedures above, the human body tends to transform. Dendrites grow back in 2 to 6 weeks with perfect compliance, faith, and prayer/intention. It is easy to interfere with this therapy through watching scary movies, taking stimulants, and wanting to prove it doesn't work.

How to treat depression? Be aware of the sense of dwelling on certain thoughts that make you feel bad. Study the feeling of separation. Apply the Scalene Light. Again, study the feelings, particularly the aloneness and separation, and how they have changed.

God immanent or God transcendent will return to the peoples' lives. They will feel more connected, either to nature or to their inspiration. Both are aspects of the experience labeled "God." Connection with nature is a result of removal of barriers to feeling the 'feminine' aspect of God. Connection with inspiration is associated with the 'male' aspect. You will find that this works vastly better if you replace the word "God" with "the transpersonal self," and "the connected self." The key word is *connected*: Connected with family, community and nature in one case, and connected with inspiration and light in the other case.

You can also use the terms joy, love, and bliss. The Scalene Light brings a sensation of Joy; a mental clarity and stimulation; *Connectedness*. The Infratonic S and Infratonic footrest bring love, which otherwise tends to be clouded by emotions, again bringing a sense of connectedness. The Vital 1 brings bliss, an awareness of the tingling life of the physical body that can be experienced through conscious presence in the physical.

The key here is the word Presence. Are you separated from your higher mental nature? Clouded by emotion from your heart and circulatory system? Separated from your nervous system through the excess of cortisol and other noise producing chemicals and methods?

This writing brings great clarity to you regarding why you have created such diverse products. You have been creating this window into the threefold world of human experience: Kapha/Vata/Pitta (doshas or forces in the ancient Indian science of Ayurveda), Jing/Qi/ Shen, body/emotions/mind, etc. Body, mind, and spirit are the results after you've removed separation using the devices. Body expands from an inanimate machine to a conscious vehicle, a remarkable creation provided by Earth to act as a vehicle, a vessel for Spirit. Emotions, when clouded, conceal a clarity of knowing and trusting which can replace the nervous system's "thinking, but not knowing," repetition of the facts presented by experts. As the emotions clear, thinking is replaced by the clarity of loving wisdom. In 'Mind,' we have viewed it as the intellect, openness to intuition, inspiration. We find ourselves to be Spiritual beings having a Physical and Mental life. The three devices are tools to facilitate the teaching of this metamorphosis.

I have left these messages pretty much as written and unedited so the reader gets the most flavor and insight from them.

9. Vital 1: Electrical Sludge, Anxiety, & Sympathetic Nervous System

"To the extent you live in your sympathetic nervous system, you are powerless."

4:00 AM

One quantifiable flavor of Etheric Sludge is the electrical sludge which causes the sympathetic nervous system to stay on high alert. Anxiety is highly disabling, shutting down the parasympathetic nervous system. This impedes healing and causes buildup of toxic chemicals in the body. The Vital 1 was originally designed to strengthen the human electric field during air travel where it works quite well. It has proven to be valuable as an environmental conditioner in a wide variety of energetically dirty environments beyond airplanes including offices, shopping centers, computers and electronic equipment, and among crowds of people or emotionally charged individuals. Our exploration of this device has brought us to a deeper understanding of Etheric Sludge. We found that people, in intense environments who feel anxiety and fear, feel immediate relief and relaxation when they hold the vital 1. What we discovered is that, while coffee and action movies tend to increase sympathetic dominance, the Vital 1, instead, moves consciousness away from the sympathetic and into the central nervous system.

The sympathetic nervous system is like a short train spur compared to the entire network of tracks in the United States. The parasympathetic is the sum total of all other neuronal consciousness in the body, including your 4th toenail, and the 4,034th hair from the 602nd freckle on your left forearm. For another analogy, the sympathetic nervous system is like a corporate or industry lobbyist seeking favor from Congress. The U.S. Congress has a responsibility to represent every citizen of the country, offering full and equal representation to all (in theory). But corporate or industry lobbyists are myopic, self-serving special interest groups, only peripherally aware of or concerned with the needs of other industries or the rest of the country or its citizenry. Each lobby has a very special interest, a strong sympathetic attachment to those they represent. Highly devoted to survival, these lobbyists will protect the interests of the corporation or industry which they represent at all costs. The same goes for the sympathetic nervous system, a couple of nerve bundles located an inch or so on either side of the spinal cord, running from the top of the thoracic vertebrae down to the 2nd lumbar vertebra. These two nerve bundles are quite slow, running at about 0.5 to 1.5 Hz, whereas the unity consciousness of the parasympathetic nervous system runs at 1.5 to 3 Hz. The special interest of the sympathetic nervous system is to identify and call attention to perceived threats to the body, to advocate for survival, to represent fear stored in the body.

It believes it protects you from wild animals that might eat you and hidden rabbit holes that might injure your ankle if you slip into them. It also believes it protects your injured body from further injury. Toward this end, the sympathetic nervous system tends to send excruciating pain signals to the brain whenever you move in a way that could possibly cause further damage. Also, it can immobilize parts of your body to "protect" you by rigidifying the tissue around the injury, whether muscle, tendon, or skin. This "protection" sometimes gives us such things as chronic muscle spasm in the neck after neck injury and permanent, hardened scar tissue to "protect" injured tissue after surgery. Both cases are examples of excess protective reaction. There are others.....

There are two sides to the sympathetic nervous system: the left side connected to the assimilative organs of digestion, the stomach, and pancreas; and the column on the right governs the part of the digestive system which is responsible for breaking down heavy oils and the cell walls of meats consumed as food. These two branches of the sympathetic nervous system have two very different perspectives. The right side destroys threats and challenges with powerful corrosive bile, a defense mechanism we interpret as anger and rage, a highly destructive method of attacking the enemy. The left side runs *to* safety. It often deals with threats by assimilating them. It focuses on protecting the body by being sure it assimilates enough sugars and carbohydrates to fuel the body. Applied to physical world threats, this aspect of the sympathetic nervous system tends to align itself with powerful individuals or groups for perceived mutual protection. From the perspective of this sympathetic nervous system, others are either loyal friends or dangerous enemies.

The sympathetic nervous system is named after this left nerve bundle, which is obsessed with sympathetic responses. "I will pretend to share your interests, and even adopt your interests, so you will consider me to be a valuable ally for mutual protection." These sympathetic bonds between people and organizations are, by nature, tenuous because the values one must adopt to become an accepted part of the group are often contrary to the best interest of the individual. Yet the perceived need for protection through assimilation is so strong that we often overlook our own needs in favor of the needs of the person or group for which we have sympathetic feelings. The reason you don't usually read about this aspect of the sympathetic nervous system is because society and the people in authority want you to be sympathetically bonded to them. If they told you that sympathetic bonding is an unconscious mechanism that is often responsible for people living in sacrifice, limitation, and fear, you might revolt against their oppression. The world has forgotten why it was named Sympathetic nervous system. Now that it is written here, you will probably read it in lots of places.

When leaders come to realize that their objectives are not sufficiently similar to ours to justify assimilation, they will "turn on us." We will experience "betrayal," a devastating experience of being cut adrift, defenseless in a dangerous world. This betrayal hurts even more because the whole foundation of this assimilating, sympathetic system is called into question. "If I trust and align myself to them, they will take care of me." "If I set aside my own needs and desires for them, they will take care of my needs and desires."

Churches, political parties, and local neighborhoods are all collectives based on this unconscious sympathetic behavior. A good example is marriage. Two people come together in mutual support and aligned interests of mutual protection, sexual fulfillment, and providing for family. Toward this sympathetic relationship, we sacrifice all those needs and wants that are in conflict with the other person's view of partnership. Often, from the beginning, both parties lose themselves in an elusive construct of mutual support. Where their individual agendas match their collective agendas couples can remain happily married for a lifetime. However, where one person wakes up to their personal needs and decides to sever the sympathetic bond, the other person feels extreme betrayal, and suffers either collapse or rage, or both.

As the sympathetic left side collapses, this previously mutually supportive and protective couple quickly switches over to the right side with anger and rage, trying to destroy the enemy, or more specifically, to destroy those interests or needs of the other person that have come between them. Once betrayal hits the assimilating side of the sympathetic nervous system, it is severely disabled. "I trusted this person, organization or group. I sacrificed everything for this person and they betrayed me. How can I ever trust anyone again?"

When we realize how the left side of our sympathetic nervous system causes us to deny our own needs and wants in favor of a pact of mutual protection and support, we can realize the importance of establishing sympathetic bonds only with those who allow and encourage us to express and pursue our own needs and wants. Armed with this knowledge, we are also prepared to begin to reconstruct our current sympathetic relationship in ways that provide us with the freedom and permission to be sure our own needs and wants can be met in the relationship. The only truly stable and mutually fulfilling relationship is one in which both parties are free to be themselves.

Political parties and religious groups often create illusory judgments and anger at outside individuals or opposing parties because anger at the enemy throws us more strongly into sympathetic nervous system dominance and strengthens our sympathetic bonds with our group, making us willing to sacrifice even more of our personal wants and needs. The United States government used 9/11 as an excuse to focus our attention on an enemy so they could create "Homeland Security," which decreases our freedom in exchange for perceived protection from terrorism.

The sympathetic nervous system is an important part of the body. However, it is myopic, focusing us either on desperation to bond and assimilate for safety, or on destroying dysfunctional bonds or rules for safety. It is an organ that embodies our fears and attempts to compensate for them. The less it dominates our lives, the more easily we can emerge into a world of abundance and true joy, love, and bliss. This pathway is to live in the spinal cord consciousness and to withdraw our awareness from the fight or flight response, the unstable balance between sympathy and rage.

Okay. What about the substance of the sludge behind anxiety? [Wherever you see a question in italics, it is me asking a question during my early morning writing.] You are correct that the sympathetic nervous system gets hyper-irritated (facilitated) and becomes quick to respond. Also, just as a focus on disaster and hopelessness facilitates the energy of depression, there is a substance for anxiety. It can best be described as stagnant electricity, electric energy on the etheric level which tends to stay in place and not move. This is the "form" stuff of fear-based thought-forms. It is created when you think scary thoughts, watch scary movies or news, or encounter scary situations. It sticks around because these thoughts are slow to dissipate. It sticks around at high concentrations in those who spend a lot of time dwelling on fear-provoking things.

In addition, this anxiety, this stagnant energy, can exist in free space. When one person tends to think about scary things a lot, the space around them gets filled with the stagnant electricity. When they leave the space, this fear remains, hanging in the air. The stagnant electricity is common in haunted houses, energetically uncomfortable hotel rooms, and offices, wherever people spend lots of time repeating their limiting thoughts and feeling, especially fearful ones.

This stuff tends to stay in place, and doesn't travel much outside of buildings because the sun's rays tend to break up or dissolve thought forms. However, in big cities, particularly those with smoggy air, the outdoor environment can become charged with electric sludge from the most coherent thought forms, those which are dwelled upon intensely for hours. This is why people often feel more anxious in cities than in the countryside.

Anxiety, or stagnant electricity, builds up the most in the limbic core of the brain and in the sympathetic nervous system. This happens because these are the places where fearful thoughts are most often experienced in the body. In the limbic core, old fearful memories can be stored since childhood and can cause headaches, migraines, and other discomforts. The chest and solar plexus are two areas which are often strongly controlled by the sympathetic nervous system. This is where a majority of the fear and anxiety are felt in the body. Also the back, from the neck down to the lumbar region tends to hold a lot of pain and tension caused by unconscious repressed fears, with fear for physical survival building up in the lumbar and fear of emotional disaster, like loss of a loved one, residing in the upper back. Stress related to supporting others or becoming overloaded with responsibilities is more emotional, and is principally stored in the neck and shoulders. These are the areas from which you are most often called to pull back your power. To the extent you live in your sympathetic nervous system, you are powerless.

How does the Vital 1 reduce anxiety? While stress is caused by the environment, anxiety is caused by limiting beliefs and internal fears. Anxiety is hyper-vigilance based on the fearful limiting belief that something bad is going to happen. This calls for high levels of cortisol which ready the nerves for action. The sympathetic nervous system runs at the frequency of low delta. Parasympathetic runs by amplifying the knowing of the heart. The Vital 1 awakens this parasympathetic function, and puts the sympathetic to sleep. *How does it do this?* By calling on the nervous system to anticipate pulses in the 2 Hz range. This is too fast for the sympathetic nervous system. The hypervigilance is focused on sensing the next unexpected pulse, and loses track of the next bad thing that is feared.

The pulses are perceived by the parasympathetic system. Vigilance shifts to parasympathetic clarity of awareness. Give someone a massage and they can't be vigilant because they are drawn into parasympathetic anticipation of the next touch or stroke. The first setting of the Vital 1 will be optimal for all applications. Setting 2 may aggravate stored patterns of separation – separation as experienced in life and separation between the sympathetic nervous system and heart. Thus, if any discomfort is sensed with setting 2, change back to setting 1. Gradually increasing the amount of time on setting 2 will gradually dissolve that separation.

It is most practical to use the Vital 1 where intermittent fears are present, such as during hot flashes or nightmares, by turning it on each time symptoms occur, then turning it off when symptoms subside. It can also be used either as needed or continuously with PTSD.

Is there a dietary cure for anxiety? Already given. No coffee/caffeine or other stimulants. A veggie diet with lots of omega 3's. Use of Vital 1 and patience. Dendrites grow back in 2 to 6 weeks with perfect compliance, faith, and prayer. It is easy to interfere with this approach through watching scary movies, hanging out with fear-mongering friends, taking stimulants such as coffee, and wanting to prove it doesn't work.

10. Mental Sludge, Worry, and Depression:
Testing the Scalene Light

"You are what you think."

What is the Scalene Light?

The Scalene Light is an environmental conditioning device with the ability to take the calming & healing effects of the Infratonic 9 and the Mobile Medic and apply those same effects to environmental spaces to lift and clear "heavy" or negative energies.

We released about 20 of these devices as prototypes to people across the United States to have them tested. Mostly they were used for clearing spaces, such as homes, hotel rooms, retail spaces, and offices from residual energies left by occupants or visitors. Those who used it in offices and retail stores found employees and visitors were more harmonious. In homes, people found that it brought greater calm and an upliftment of mood. In one case, the pets started behaving with far fewer dysfunctional habits. In another case, a woman lived with her brother who worked as a counselor of sex offenders in a local jail. She found that, before, she could not enter his home office without feeling intense fear. Once she cleaned the space with the Scalene Light, she could enter the room comfortably. The whole home became more comfortable. The Scalene Light is effective at dissolving or erasing the mental structure that causes the sensation of discomfort. I have found that this device is useful, in addition, for people who are suffering from grief or depression in that it temporarily dissolves these patterns in the mental field of the room surrounding the person. This gives the person about two days to experience what it feels like to be free of these feelings, and to observe which of their thoughts, emotions, and interactions cause these feelings of grief or depression to start coming back. Thus it acts as a biofeedback tool. One of the main uses of the Scalene Light has been in interpersonal conflict. Clearing a space in which there has been habitual conflict has generally caused that conflict to diminish.

How does it work?

A stationary linear electric vector potential is generated by a stationary rotating magnetic field, a circumferentially magnetized cylinder. This combination entrains photons, similar to a laser. The Scalene Light softens thought forms and electrical stagnation, allowing them to dissipate. It works by inducing a ray that temporarily suspends the thought side of thought forms so the astral (magnetic plane) and physical (electrical plane) aspects are free to dissolve. The double heart frequency induced onto the ray transmutes low frequency, heavy thoughts, to love/wisdom, (a higher frequency). The LEDs produce photons that ride between the astral and the mental planes (penetrating opaque materials like clothing and walls), and temporarily decouples the thought from the form, further softening the thought form for faster dissipation.

Who can benefit from using this technology?

Though not everyone is aware of negative or dense energies, everyone can be affected by them. Therefore, this technology can benefit everyone. Animals are especially sensitive to the energy in their environment, so this is beneficial to them as well.

Where can it be used?

The Scalene Light can be used anywhere that has negative or dense energies, as well as places where many anxious people (or animals) may pass through, or where there might be interpersonal conflict.

- Hospitals, Emergency rooms, Surgical rooms, Urgent Care facilities, Recovery rooms, Medical offices, Dental offices, Rehabilitation facilities, Therapist offices
- Hotel rooms, conference rooms, bars
- Business offices
- Government offices
- Mega-retail stores
- Chain restaurants
- Classrooms
- Daycare facilities
- Foster care facilities
- Churches
- Dog pounds, pet stores and animal rescue facilities

When should it be used?

Many people can sense or feel congestion in a space, sometimes feeling it as darkness or heaviness, and sometimes as sadness, anger, or other heavy emotions. It is important to use the Scalene Light in conjunction with this awareness so that specific locations of congestion can be identified, and increased lightness and elevation of mood in the space can be verified.

How should it be used?

There are 2 timer settings on the Scalene Light. One setting is for 10 minutes, and the other setting is for 40 minutes. Total daily usage should not exceed 40 minutes. A third timer setting causes the device to operate for 3 minutes every 3 hours, and is used for continuous clearing of a space where there are recurring sources of congestion such as shops and therapy rooms.

There are 3 intensity settings on the Scalene Light: Clear, Soothe and Penetrate. It is recommended to begin a space clearing by using the Clear setting and finishing with the Soothe setting. If the Clear setting is not penetrating the congestion adequately or breaking up the local etheric congestion, you will want to use the Penetrate setting, then follow with Clear and Soothe.

The Clear setting is generally the best signal to use to clear spaces. Apply from a few feet away to several feet away to break up stagnant energies in the field, similar to breaking rocks into sand.

The Soothe setting is the most harmonizing and it is best to finish with a minute or so using this setting from a distance of 10 to 20 feet or more to gently dissipate and smooth the residual stagnation in the field to leave it feeling smooth and harmonized. This helps a cleared field to stay cleared, to integrate the shift and make the harmonized field last longer.

The Penetrate setting is the most disruptive signal and is to be used sparingly. Apply it locally where the Clear setting was ineffective at breaking up etheric congestion.

The Scalene Light is best used a little at a time, to allow thought forms to dissipate and to allow the environmental space to "heal" or to become accustomed to the cleaner condition. Like people, spaces may not be ready to release everything at one time. An initial clearing will remove a lot of Etheric Sludge, but trying to clear everything at once can create unnecessary stress, thus, in some ways, increasing the Etheric Sludge in the space. If there are situations where someone has been especially upset or angry, a clearing immediately after their departure from the space is very beneficial. The Scalene Light should be stopped as soon as the space is relatively clear.

The Mobile Medic or the Infratonic is very helpful when used in conjunction with the Scalene Light because they work on the astral/magnetic and physical/acoustical planes, breaking up, harmonizing and integrating the desired patterns and inducing an alpha state which transmutes desires like "I want" and "I need control," toward "I want to help and cooperate." In addition, the Mobile Medic/Infratonic can act similar to a water spray to help cleanse the area and soften stubborn thought forms. An intention is clearly helpful, such as "seeing the flow of etheric water from the Infratonic and washing the stagnation away, sweeping away etheric debris released by the Scalene Light."

Comparing the Scalene Light with Infratonic Technology: The Scalene Light dissolves mental debris in the environment. It clears heaviness, allowing expansion. It allows people to feel more comfortable in their environment so they are more able to feel the comfort and safety they need to relax and allow the healing process. The Infratonic harmonizes the heart and emotions.

What are the precautions regarding the use of the Scalene Light?

It is not intended for treatment of humans. It tends to have a subtle effect of clearing energy blocks in people, so it can tend to bring emotional stuff to the surface. It is for this reason that we recommend that it not be applied to people, and not be left in one position directed at a person.

Some people have expressed a desire to apply it to people. We recommend that the Scalene Light be applied instead to the spaces that people occupy. Clearing a space clears the thought forms that people have released into the space, and people find that they immediately become much more comfortable in the spaces, can relax and expand, and are more able to release their own stuff. Do not apply the Scalene Light near a person who is uncomfortable with it.

Examples of Applications:

Offices of 30 to 50 people: I have had two opportunities to apply the Scalene Light to offices, one accounting office, and one research facility. Both were quite similar. Each visit was prompted by complaints of conflict in the workplace, and in both cases, the employee who requested the Scalene Light could easily feel congested energies. As we went from room to room, they would feel where the energies felt the most congested and we would apply the Scalene Light until the energy became relatively clear. Some rooms or work spaces were much more congested than others, and these sensitive people could feel different emotions associated with the congested energies, whether sadness, stress, depression, anger, etc. We found that rooms could be cleared through walls, doors, and windows, and a line-up of 5 cubbyholes, (partitioned offices) could be cleared at once by aiming the Scalene Light in a direction that passed through all five workspaces at once. The Scalene Light would also project 50 feet along the edge of a room where the wall meets the ceiling, making it easy to clear congestion that tends to collect along these edges. In both facilities the initial result over the first few days was a more harmonious atmosphere with less conflict, and in both cases within a week hidden conflicts came to the surface, were addressed and resolved. The Scalene Light treatment was repeated after one week with a smaller amount of sludge to clear and increased gains in interpersonal harmony. Without more Scalene Light treatment, over the next few months, there was some return of the heaviness, though much of the interpersonal conflict stayed away.

Coffee Houses: I experimented with coffee houses like Starbucks. The buildup of sludge from constant visitors was greatly reduced within 5 minutes of directing the Scalene Light, set on Clear, around the room. The next day it was still pretty clean. After 4 days much of the sludge had come back. There was one incident where a server was having a bad day, gloomy and distant. Within 10 minutes after the clearing, she had popped back to her normal cheery self.

Miscarriage/Depression: One of my employees had a miscarriage and was missing days of work because of depression. She would come in for an hour every few days to get the essentials done. Once when she was in the office she agreed to have her office and the field around her cleaned with the Scalene Light. After 5 minutes of directing the Light around the room and around her field on Clear, then a couple of minutes on Soothe from about 10 feet away, she said "Strange, I feel fine." She was fine the rest of that afternoon and the next day, and worked her normal hours. Then during the weekend, she visited her mom and got back into the energy of depression. She had apparently rebuilt the depression in her field because she missed Monday and came in for an hour on Tuesday. I found her in the office, again depressed. I said I thought there must be some aspect of the lost child she was hanging onto and that she might need to find and release that attachment before she could be free of the returning depression. I asked if she was ready to let go of it. She said "No. I'm not ready." About 15 minutes later, after we had gone over some accounting work, she said, "I think I'm ready now." So we got out the Scalene Light again did a brief clearing of her field and the room again from a distance of several feet away, then started discussing what her attachment might be. After a few minutes of discussion, we cleared the field again. She said good bye to the lost child, and was free of the depression from then on.

This illustrates the concept that we create our moods with our thoughts, desires and beliefs, and can rebuild a problem after it has been cleared from our field by returning to a place where we felt it strongly, or by repeating the mental/emotional processes that created it in the first place. This may explain why it is so common that, after a massage or chiropractic treatment, the patient is often free from pain for about two days. Then the pain and tension often return and the patient needs another treatment. While we can temporarily improve comfort by clearing the field or having others clear it for us, people tend to rebuild the thought form by continuing to hold thoughts, emotions, beliefs, and judgments, particularly self- judgments like "I'm not good enough," or "It was all my fault." Also, we can build up these patterns in the places we occupy, like a bedroom, office space, or reading chair. While a professional might clear energy in and around us, when we return home, or to the office, the memory of the field will tend to induce the dysfunction back into our field.

I have come across several other people who have had concerns, emotional overwhelm, and other "sludge overload" conditions of various sorts. I have found that simply by circulating the Scalene Light around these people's fields from 10 feet away, their concerns temporarily vanished. This is a very useful educational method if the person is first encouraged to experience what it feels like to have this "Etheric Sludge overload" condition, then afterwards, to feel the freedom, elevated mood, and expansion that come after the Scalene Light clears the sludge.

There is also an important third step. If they are reminded to be aware of what they are thinking or dwelling on over the next two days, and to be aware of when the "sludge overload" condition begins to return, they are likely, through introspection, to discover what they were thinking, worrying, or stewing over when they recreated the thought forms in their fields and recreated their symptoms. This introspective awareness appears to be a most important method of healing.

Objective Verification:

The Gas Discharge Visualization (GDV) device, which is a computerized Kirlian camera developed in Russia, has an attachment called Sputnik, which measures and quantifies the etheric congestion of the atmosphere. This device was used in conjunction with the Scalene Light in clearing a house to discover whether the GDV Sputnik could quantify the change in a space. This experiment was also conducted with an energetically sensitive person who examined the room, identifying areas of high and lower etheric congestion before and after clearing the space with the Scalene Light. It was found that applying the Scalene Light to a congested space caused the GDV image to become less bright and larger in diameter. This suggests that Etheric Sludge causes the space or the air to become more constricted and less conductive, such that electrons ionizing the air have more trouble moving through it. Thus electrons release their energy of ionization over a shorter distance and produce brighter light over this smaller distance. Upon application of the Scalene Light it appears that the space (or the air) becomes more relaxed, allowing electrons to flow more easily, creating less luminance over a larger area.

Through this GDV method it was found that the Scalene Light can be directed to clear local areas of Etheric Sludge in a room, leaving the rest of the room only slightly cleansed. In addition, the GDV provided objective evidence that this Etheric Sludge tends to accumulate more in corners than in the middle of rooms, though also collects in those places where people have spent a lot of time in intense emotional/mental states.

DNA is used to quantify Etheric Sludge:

What is the nature of this atmospheric stagnation which allows it to be stored in the space of a room? Russian researcher Peter P. Gariaev conducted compelling research which offers insight into information stored in space, and indirectly, atmospheric congestion. In his 2006 groundbreaking paper "Crisis in Life Sciences, the Wave Genetic Response," he shows that a sample of DNA in a test tube can absorb light from a laser. Then when the laser light is stopped, the DNA sample starts emitting light. Then, when the test tube is removed, a most unexpected thing happens. The location in space where the test tube had been continues to emit a characteristic pattern of light, gradually fading over the next month. This provides objective evidence that the space, whether the air or a substance that stands behind the air, can receive, store and send information.

To determine whether the light was stored in the air, or in the space independent of the air, he directed some compressed nitrogen at the location where the phantom emission was being measured. The light emission immediately stopped, but came back in about 5 minutes. This remarkable finding suggests that the air is what emits the measured light, but that some quality of the space, independent of the air, stores the etheric congestion and reorganizes the air such that it can start emitting the measured light again. This seems to be solid evidence that empty space contains a substance that can store energy and information. A principal difference between dry nitrogen and compressed air is that air contains oxygen and water vapor, which, unlike nitrogen, have significant magnetic qualities. These gases and magnetic fields may play a role in how this stored energy in the space structures the air, and why dry nitrogen is able to temporarily disrupt this structure. It appears that this phenomenon of DNA information being stored in a space may be the same phenomenon as atmospheric congestion, the emotion, or heaviness that gets stored in a space.

Etheric Sludge, also known as bad Qi, dirty energy, or stagnant air, has been around for thousands of years, along with methods of clearing it, such as opening windows, burning sage, and sounds, like clapping hands, bells and singing. The Scalene Light offers an objective tool that

157

effectively clears Etheric Sludge from spaces, offering people the opportunity to experience the difference between a dirty space and a clean space. Even those who are not energetically sensitive can experience a big difference in their emotional/mental state before and after Etheric Sludge is cleared, providing insight into the nature of our mental/emotional experiences, showing us how the space we occupy influences our moods and pains, and how we influence the space we occupy with our thoughts and emotions.

The Scalene Light and How We Create Imprisoning Thought Forms

The theme of this book is releasing limitation to expand into abundance. Here we take a look at how we create limitation in the first place.

Each time we create thoughts or words, they go out into our fields, our auras, and stay there for a while, and then fade away. If we create similar thoughts frequently, they get stronger and stronger in our fields as they gain physicality. As they get stronger we can start to perceive them more and more strongly as part of our reality, even if they are not coherent with the consensus reality around us. Thus, if we repeat something that is false often enough, we will begin to perceive that it is true in the world around us. This is the source of our denying or not perceiving the truth when it is right in front of us.

Further, as we build up these thought forms in our fields, these forms begin to influence our physical, cellular body, and whereas our cells have their healthy nature, when they are influenced by strong, built up thought forms that are not coherent with our physical being we are, in essence, warping our field and creating abnormalities in our physical, cellular structure. If we believe strongly enough that we are weak, we become weak. If we believe we are blind or do not want to see things, our vision will start to fade. If we believe we are getting old, our aging will accelerate.

The physical consciousness is the collective consciousness of the physical body. As we go through life, we come upon beliefs about who

158

we are, and what we are and are not capable of, and gradually make these thought forms so strong that we perceive them to be real. In this way, our physical consciousness takes on beliefs that are not necessarily coherent with the world around us. We limit ourselves and become blind to possibilities. Through meditation, if we can become silent in thought and word for long enough, these thought forms will fade enough that we can begin to see through our own illusions, and will begin to see the consensus reality around us more clearly. Some people see the ego as the enemy and try to destroy it, but it is just the thought forms that we keep re-enforcing with our words and thoughts which we have come to believe that they are the problem. As we open to the possibility that we are far more than we can imagine, and that we are really not anything like we believe we are, the old illusions will gradually dissolve and we will see clearly. The key, again, is releasing limitation to move into abundance.

"Consensus reality" is not the true way the earth is, but rather the creation of a group of people. Just as we can fill our individual fields with false beliefs, all of humanity is constantly throwing around thoughts and words to create a consensus reality that is often, in many ways, false. For instance, it was long believed that the world was flat, and governments are constantly rewriting history through PR and the media. Currently, consciousness is changing rapidly, but the world of thought forms within which humanity lives is still thick with the old illusions. If humanity is to break out of the past and expand with this new consciousness, we must individually shake off our own illusions and be open to possibilities as they emerge and reveal themselves.

The Scalene Light is a tool to help soften the thought forms that surround us. It dissolves what some people perceive as darkness or heaviness in the aura, and leaves people feeling lighter and freer. It is just a tool, and does not directly affect what we say and think. As our cloud of illusions begins to fade, we find our tendency to dwell on unnecessary thoughts or repeat counterproductive phrases begins to fade as well. With the Scalene Light, the role of intention is extremely important. For best effect, we must intend to release old baggage in our fields, and open to new insights about ourselves. We must also, to the best of our ability, refrain from re-enforcing any repeating thoughts or concepts. It is hard to "not think" or "not speak idle thoughts" or "not gossip" or "not judge" but it is precisely these acts of intentional restraint that help us to keep the field clean between uses of the Scalene Light. Intentional restraint is the true key toward expanded consciousness. The Scalene Light is just a tool that accelerates and facilitates the process by dissolving the thickest of the thought forms so we can see more deeply into the world of illusion we have built for ourselves.

Parallels to the Concept of Imprisoning Thought Forms:

Darkness: The Scalene Light is often perceived as clearing darkness in the aura. Darkness is the densification of light so, as the Scalene Light is applied using the Penetrate or Clear setting, it tends to expand areas of darkness so that they are more diffuse and less dark. Further use of the Scalene Light on the Soothe or Clear setting and at a distance of 10 to 30 feet away causes these diffuse areas to become yet more diffuse and to dissipate. This darkness is a visual for buried, repressed, or otherwise densified thought forms.

Energy Cyst: The Scalene Light seems effective on Energy Cysts, which are areas of the body which are encased in a darkness that separates areas of the body from the rest of the body. They are often generally spherical, though can be of any shape or size. Darkness around individual cells can isolate those cells from the body, and darkness in the aura that surrounds the person can isolate that person from others or from Spirit, resulting in a feeling of separation, numbness, depression, or non-feeling. At any level, the Energy Cyst causes the cell, tissue, or person to lose touch with their true nature. The Scalene Light dissolves this energy cyst, restoring oneness with the surroundings, and helps individuals to remember, and more embody, their true nature.

Cellular Constriction: Thought forms, or the Etheric Sludge around cells, tend to clog the cellular structure, and much like molasses in the intracellular fluid, impedes the movement of fluids and materials into and around the cells. Use of the Scalene Light might, by dissipating residual thought forms, vitalize sluggish tissue and increase the rate at which this tissue receives nutrients and releases metabolic waste and toxins.

The Scalene Light brings up old stuff. Old, buried issues tend to start expanding and dissipating with the use of the Scalene Light. While much of this old stuff is stuff we are ready to let go of and which dissipates readily, some of this stuff still has a charge that is closely linked with a whole network of related old stuff. Often, with the use of the Scalene Light, we will experience situations in our outside lives more acutely, and perceive our role in creating issues more clearly. Sometimes we will remember related childhood situations and see how we continue to employ old survival approaches in our current lives. At other times, we will have a progression of discomforts as successive globs of related stuff expand and dissipate. While we don't usually need to experience old pains or relive childhood trauma it is valuable to intend that all issues that are ready to surface, present their lessons and dissipate, and that we be left with more effective or appropriate interaction skills. Since we are extremely complex from the standpoint of beliefs, desires, hidden agendas, and buried trauma, issues will emerge when our whole being is ready to release them. This by necessity, is a progressive process, and the more clear we are of cloudiness in our fields and the more ready we are to learn new ways to see and interact with the world, the faster these issues will emerge, share their lessons and dissipate. With patience, our awareness will expand more and more as the darkness within dissipates.

Quantifying Sludge

Once we have a measure, we can start to quantify spaces and "clear" or "enhance" spaces through methods such as space clearing, color, or sound then measure again. We can compare this to perceived changes in quality of life of sensitive people who occupy these spaces.

As an example, in a hotel room which customers have complained feels creepy or uncomfortable, we can start by having sensitive people evaluate the room, particularly identifying what corners or areas have the greatest disturbance. We can also quantify using electronic means, clear the space, and quantify again to validate that the intensity of readings is reduced. Finally we can have sensitive people evaluate again. This is a valuable service to hotels in that they have happier guests and fewer complaints.

This method might be applied to hospital rooms, where patients frequently have their energy fields torn open and go through intense emotional experiences, impressing the space with intense energies and leaving the room feeling uncomfortable. Clearing these spaces improves the patient experience, the experience of doctors and nurses, and probably outcomes for patients and long term health for employees.

This method might also be applied to a work place. As people sit at their desk on a daily basis, their work spaces become cluttered with their thoughts and emotions. This stored information begins to reinforce their thoughts and feelings. Where these thoughts and emotions are directed toward effective and innovative work, these spaces become empowered and clear. However, where employees focus on problems from home or interpersonal issues in the workplace, the space becomes cluttered with disruptive patterns.

The Scalene Light is truly a breakthrough, not just as a device to clear old, thick thought forms in spaces, but also as a teaching tool to bring into science the oft quoted adage, "you are what you think". As we, as a society, learn that we have a choice between literally sitting in our own mental filth and rising above the sludge and leaving limitation behind, we truly open the door to abundance.

11. Inverse Delphi Method: Scalene Light

Brain Mind and Healing Program

The energy field within which we live influences our lives and connects us to all that is. Indigenous cultures around the world all acknowledge this energy and have developed a wide variety of technologies to cleanse, heal, and enhance it. And yet, we, in our modern scientific world, know so little about it. 20th century physics denied that this energy field even exists. We influence it with our thoughts, feelings, and words. In addition, other things influence it like lights, sounds, scents, plants and animals, electromagnetic energies, and a wide variety of therapies. We can modify the content of this field with our own focused thoughts, words, and intentions.

The challenge in studying this field is defining some scientifically valid frame of reference by which this field can be identified. This is most challenging in that most people don't have any perception of this field, and don't even have a belief that it exists. Those who do perceive this field seem to perceive different aspects or layers. Even those who agree about what level or nature that they are perceiving, will often have conflicting perceptions. It's simple to say, "Well, it's subjective." All that means is that we don't think we can study it scientifically.

How do we measure this field scientifically? We can look for consistencies among sensitive people as they describe the "qualities" of the field, but that doesn't fit the old scientific paradigm. What would be better would be to use electronic equipment to "quantify," this existence. The field itself appears to be imperceptible to most equipment. When people perceive these noiseless fields as calm and peaceful, electronic equipment measures zero. Then, as we get into field activity that is heavy or frenzied, electronic equipment will measure increased noise. Thus, while this is a crude measure, it can be correlated to subjective impressions of sensitive observers.

Chinese scientific research into Qigong has found that especially talented Qigong healers could induce changes in most sensitive equipment, apparently with a combination of intention and "emitted Qi." The fact that emitted Qi can either increase *or* decrease the growth rate of cancer or the decay rate of radioactive substances shows that it is not Qi itself, but rather, the result of the emitted Qi intervention. From a scientific perspective this Qi or energy field is very elusive. I worked for years trying to develop sensors that would measure this stuff, and measured a variety of rhythms in the body, some not previously measured by science. However, this proves nothing about the substance of this energy, only that some aspects of the body vibrate according to the measured frequency pattern.

Thus, I changed my plan from measuring Qi to building therapy devices to see which devices would influence the body, and what their influence was. Then, we would know that a specific electromechanical device influences the body in a particular way. This will tell us that whatever substance in which the device is producing signals, is conducting those signals, and those signals, induced in that substance, are causing particular effects. Thus this method seeks to isolate substances of the vital field and influence vibration upon those substances. By restricting the resonance of the devices to what I believed to be single forms of substance, we could get an idea of which substances are important to all different aspects of the body's field.

Then I came to a bigger problem. After building the devices, how do I test them? How do I determine how they influence the body, what they are useful for, and how to optimize them? For this, I employed a research methodology which I had read about in psychological literature, the "Inverse Delphi Method." The idea is that you can arrive at truth by asking a bunch of experts a question, then summarizing their answers, and asking them all the same question while providing the summary of the group opinion to them. Most of them will eventually agree on an answer. Well, I didn't really use this method because I discovered long ago that averaging the opinions of a group of people in search of new findings is often useless because of the dilution induced by the committee effect.

It was while working at the Gas Company that I learned that this method was useless. I was taking a management training course, and they were having us do an exercise in which they had us answer a bunch of difficult questions about how to manage difficult employees. I scored a 72 on it. Then they had us get together in groups to discuss our answers and come up with a consensus. The idea was to show us that five heads are better than one. However, in my group, after much discussion and compromise, we achieved a score of 47 on the same questions. The problem was, of course, that I was not forceful in presenting my solutions, and others with a more dominant nature were more persuasive that I was. I didn't really care. I just threw out my answer, and if nobody was interested in it, that was ok with me. I have always been quite good at seeing multiple sides of a situation or an argument, so was quite willing to consider each person's position as a possibility. Where most groups in this management class learned that working together provided better answers, I learned that people are more interested in persuading other people than in arriving at the best answer; so much for the Inverse Delphi Method.

I developed my own version of the Inverse Delphi Method (I liked the name). I would build a device that popped into my head, apply it to some people and see whether it seemed to do anything, then, if it did something, I would question lots of people, mostly psychics, energetically sensitive people, and people who do applied kinesiology. I would ask them how the device affects them and what it does, how it works, and what it's good for. Based on the commonalities in their responses, I would create theories, build new devices or develop new signals, try them out again on a few test subjects, then go back to the intuitives to ask more questions.

I would like to illustrate this by telling you how I developed the Scalene Light. I was studying the Infratonic device and pondering how it worked. I saw that you could take two electrodes and apply a signal to the body. This created an electric field across the body. Then you could take the same electrical signal and apply it to a coil of wire, convoluting the wire to create a coil. This would create a magnetic field that you could induce into the body. In so doing, we advanced from the physical field (electrical) to the emotional field (magnetic). I reasoned that we might do this another time and reach the mental field. By convoluting the coil of wire into a toroidally wound coil, it seemed we might create a device that resonates in the third substance, the mental substance.

So I built one, an early model of the Scalene Light, and tried it out. I found out that it seemed to relieve some pains that had been difficult to relieve. I then went to intuitives and started asking "What does it do? How does it work?"

My intention at that point was to have this be a therapy device. Now I have discovered that is more effective as an environmental conditioning device, for hotel rooms, offices, homes, wherever energies get dense and need to be cleansed. It works like high powered Sage only it is cleaner, doesn't create smoke, and is far more effective at clearing out heavy energies in a space. Thus the Scalene Light which we offer is an environmental conditioning device. The feedback below from intuitives, while offering great insight into its effects as an environmental clearing device are couched in terms of treating the human energy field as opposed to the energy field of a room. Nonetheless it provides an entertaining exploration into the use of my version of the Inverse Delphi Method toward researching strange and exotic equipment:

What does the Scalene Light Do?

I started by asking a friend who was visiting who worked as a hospital nurse at night and a psychic reader by day. She described its effect as opening a portal to the descending of a new energy grid for Earth. That was cool, but vague and New Age.

I then asked a woman I met in Sedona, the girlfriend of a capable energy worker. She described it as causing the thought forms surrounding the body to start vibrating more and more erratically, then abruptly convolute into higher order, simpler geometric shapes which had more dimensions than the original thought forms, like three dimensional shapes that turn into 4 dimensional shapes. She saw a great simplification of the energy field of the body. That sounded like chaos therapy and supported the idea that this might be engaging the mental substance.

Dr Peebles and the Field of the Heart

I arranged for a channeling from "Dr Peebles" a being with a Scottish accent who is channeled by many people around the world. He offered the following:

"It is a point of focus that helps you center. It harmonizes you, mind, body, and spirit, keeping you focused and balanced, feeling relaxed, allowing you to be in a vulnerable state without feeling the vulnerability, without feeling the fear. So what it does, really, is it dissipates any negative influences around you. It can block out voice patterns as well as hormonal patterns, energetic patterns of others who might try to influence you when you are striving to remained focused on your journey, when you are striving to stay focused in your heart. It's not doing the work of staying in your heart. You're doing it, but let's say you are doing it in a somewhat controlled environment, which is what the Scalene Light is providing for you. It's wonderful. It's a good harmonious energy. It's not harmful. It's very gentle. It's also very fragile. It can dissipate easily. The slightest bit of doubt can influence it. It's a very beautiful energy.

"It's similar to if you walk into a room and someone else walks in who has the heart of Christ. They're not trying to manifest anything. They're not trying to make you feel anything in particular, but you can't help it. You are in their presence and you say, "Oh Goodness, I just feel so suddenly safe, relaxed." You would just become that which you are seeking and that's really what it's about here. So the presence of this Scalene Light is like the presence of Christ, really.

"It is not necessarily going to be popular because people mostly want to be fixed. They don't want to do the work. They will have to. In this energy, once they are feeling alright, it's "Okay, what are you going to do next?" Now for people to be healed, they will say "I want healing here." But in truth, my dear friend, there are plenty of people who are

167

terrified of having a whole healing because then they would have to take responsibility for their journey.

"Best instructions to accompany the Scalene Light: Relax, Release, and Surrender. And allow for it to do its work. You don't have to do anything. This is a tough thing for people. That's why it's not going to be your most popular product.

"It would provide a controlled environment, a sensation of well-being in which one can then do the inner work without the confusion, without the conflict of emotions and fear and anticipation, and expectations. It is a place, again, that provides the individual with the security of being able to be vulnerable without having to feel vulnerable.

"Scalene Light Physical mechanisms: Does the Scalene Light disperse stagnant electricity? Yes it does. – That's always very healing in and of itself. You can do it yourself just by allowing static electricity to drain out your palms toward the floor. For those who don't like to think in terms of anything that would be considered New Age or what have you, who like technologies, the Scalene Light will work for them.

"Can the Scalene Light be described as tuning the parts of the body to which it is applied to the vibration of the heart? Yes, alright. If you want to describe it like that, it's alright. It doesn't tune. It is what it is and the individual winds up harmonizing self with the device, but if you want to put it that way, it's not a lie. It's a good way to describe it.

"What sort of damage can it cause if people use it too much? We don't see that there is any damage to be done other than addiction because if one is feeling good in its presence, and doesn't feel good away, then one is not doing the proper work with it and so becomes addicted to the sensation of well-being that it might provide, so the point of the Scalene Light is to use it for a very short period of time, learning how to harmonize one's self without having to use the Scalene Light. Can you do it in the controlled environment, then go out in public into a chaotic situation and still feel that centering and that peace, that balance?

"If people, instead of surrendering to it, are filled with fear and worry, can it open them more to fear and worry? No. Because they are

in a controlled environment and feeling well, so what happens is, one is allowed to be vulnerable without feeling the vulnerability, so you can sit in this frequency and be afraid but not feel the fear, because what you're really doing is being driven back to the source. The fear exists, the concerns exist, and the pain exists. It is all there because it just is what it is. But you are no longer fighting it. You are surrendering to a greater truth, which is love. And that's really the bottom line as to what is occurring here. That's not going to improve your bottom line however, because it's ethereal.

"You can't say this technology is doing anything besides providing a controlled environment of contentment in which one can grow and learn and increase awareness of self. And that's really what it does. It is a wonderful environment."

Pleidians on the Scalene Light

I then arranged for a channeling with a woman who channels the Pleidians. Here's my summary of what they said:

The Scalene Light clears out toxins like metals and artificial chemicals that are locked in the cells. Our cells attract toxins because of lower frequencies or fear frequencies. The Scalene Light works by raising the frequency of the cells so their attraction for these toxins disappears and your body processes the toxins out. The Scalene Light also releases things in the emotional level. If you don't do the emotional clearing, the physical work will not hold. This should be pointed out to users of the Scalene Light. The biggest thing the Scalene Light can do is show you what a better frequency feels like. It can temporarily raise you out of the frequency of chronic pain so you can feel it. Sometimes that's enough to allow you that extra boost to do the other work.

An SR (Scalar Ray) Reading by Joel Bruce Wallach, Clairvoyant

This was an extraordinary series of readings providing tremendous detail and insight into the operation of the Scalene Light and its interaction with congestion in fields and in spaces around organs, cells and atoms. For these readings we used a preliminary prototype, called the Scalar Ray, hence all the references to the SR. The early prototypes were more aggressive and intrusive, and would have a bit of a traumatizing effect on spaces. We made many changes through several iterations to arrive at the Scalene Light, mostly in the direction of making it more harmonious and friendly, more compatible with the spaces and fields we live in. The only "feature" we eliminated was its tendency to "eject entities," as this is best if it is by the conscious choice and request of the individual.

Joel: [Holding the SR] "Modulating a Scalar wave. The energetic structure, being a tube, implies that higher consciousness will be flowing through because, it's spinning, which is making it more conscious, and because the tube is going to start to wake up with, one hopes, higher order fields. So, let's tune into it because that was just a mental description. It is both directional and non-directional.

"So, now I'm placing it in front of my heart and the LED is off. It's calming. However, the energy it is sending in is starting to go in almost like an intelligent wormhole, so we will track it and see what it does. It's starting to impart some higher order healing information that would not ordinarily be in a stationary magnet.

"I'm going to look at it close up. I have a diagram here of atoms and molecules in symbolic form. I'm going to look in to see what's happening at each of those levels.

"At the atomic level, I'm putting the device on the solar plexus. The electrons around the atom seem to be more slippery so there's less friction in them. The resonance between the fields or shells around the nucleus of the atom seems in more contact with the nucleus when the device is against the body. It definitely goes through layers of tissue.

170

"We'll turn it off for a little bit. It's continuing to permeate through the body. It's continuing to descend through layers closer to the core and the spin is starting to send something energetic. The energy feels like, qi or prana, which is starting to move through and disperse it through the body from the area it was placed. Also, it seems to improve circulation. Qi or "Life Force Energy" and circulation are closely related.

"Let me psychically ask the atoms what they've learned and if they're still learning anything or if they have any insights – the device is still off and the electrons are saying, "This reminds us of something. It helps let us recognize something in which we have forgotten." So I'll ask them, "What do you remember or recognize?" And they say, "This was the way we were in the original, perhaps pure pre-human form." So there is something of the eternal that they have been reminded of.

"I'm still feeling it permeate through so I'll track that and then track what is happening at different physiological levels:

Molecular level: There is more coherence: "How does this help the body to respond to toxins? It might help the body strengthen itself so it can address toxins, but this device does not, per-se, address toxins. Somehow, the molecules are expressing that they do not feel so constricted. "If we continued to use this device, what would happen?" "It would help the molecules remember prior order or coherence." That almost sounds like transmutation. If you have some vibration in you that is not in alignment with yourself, the device would help you to remember your true self signature, which we assume could only be a purest of intentions.

Cellular level: The cells are saying, "Let's try this again." Now we'll put the device on the feet. "What's happening?" "It's like a Jacuzzi" "How so?" They're referring to a scrub brush.

"At this point, we have the power on and the LED off. This increases the field strength of the cells. They seem to be throwing off an energy field that looks as if it is dissolving. Maybe just like oxidation. This implies that this functions like an antioxidant at a certain energetic level. It seems that, in the area in and around the cell, little flecks of debris that didn't belong there are starting to dissolve because the field strength of the cell is stronger. So it is somehow differentiating between itself and the debris, whatever they might be.

"Also, it is giving the mitochondria a charge similar to plugging it into a circuit, in a healthy way. So we're talking about a healthy form of bioelectricity. "Anything else on cells?" "Yes Joel, there is more coherence. The nucleus of the cell is starting to repeat the rotation. It seems to enjoy that and it seems to remind itself of a higher expression of its true self.

"Now, let's turn the device off again. "What happens to the aura of the cell when it experiences the power on and the LED off?" Before the device was applied, I see that the field around each cell was somewhat weaker, which would make them vulnerable to invasion of debris that would not be in a normal cell. It seems as though the cellular membrane had been weakened. So it is spontaneously pushing out debris that would not normally be in healthy cells.

"Bone/muscle/nerve level: It brings to the surface, distortions and traumas or emotions and thoughts that have been trapped. This means that it is removing those things that might be held in emotional mental complexes that are usually held in the cells. This could be the cause of a recurring or terminal health issue.

"Looking into the bone/muscle/nerve structure, what happens to the coherence and correct functioning of the structure?" It seems as if it's coming out of a slight daze where it might have forgotten its correct coherence.

"The mental and emotional patterns are starting to release, so the physical self, maybe not quite in present time, is remembering a traumatic event or experience. The emotional trauma or "the past" is starting to come out of the body, and is metamorphosing into present time. It is remembering here and now, living in the moment.

"If you [bone/muscle/nerve] receive some of the vibration of this device on a daily basis, what would occur?" The structure seems to be saying, "It helps me to remember who I Am," which I suppose means that "I forgot who I was." In other words, the bone/muscle/nerve's identity is being aligned with the natural order of all things; Nature. It doesn't define things in emotional terms or false beliefs, but rather says, "Oh, I was confused, and now I'm getting more clear about who I Am." So, it is a fairly simple system of perception that seems to be very relevant to its level"

"Etheric Template: "Is the tissue getting closer to the etheric template, or the etheric template getting closer to something more?" When we hear the structure saying that it's remembering something, it's simply saying, "I remember." "Etheric template, are you feeling anything either directly or indirectly as a result of your experience with the device?" "Yes! Plenty." "It transmits a picture. It shows that "Natural Order," is being improved at all levels. So, the etheric template is showing us that there are different octaves even within the ether. The setting we have done so far without the LED's is working on some of the lower octaves of the ether. We can turn the device back on with the LED's "on."

"What do the LED's contribute, if anything, to the etheric template?" They're working on some of the octaves of the subtle etheric and emotional bodies. It is not however, working on the mental or spiritual levels.

"I will go back to the atom/molecule/cellular level and ask at the mental and spiritual levels: "What do you experience with the LED's combined with the magnets?" It seems as though it is slicing through bad influence. It seems as if any distortion of any kind ... the spinning LED's can melt through negative influence more effectively than the *rotating magnets* alone. Also, there's a slight massage effect that wasn't there before using the magnets. It's almost as if there is a tiny inner spiraling vortex spinning like a "roto-rooter," that is working at the atomic, molecular, and cellular levels. So, there is vortex activity at a very micro level. I feel as though it just puts me at the next level of calm.

"I will imagine someone in a third world country with a tubercular condition for these maladies. We ask the body, "What does the SR do for you if you are afflicted with a bacteria or a virus that is fast growing and affecting your respiratory system?" If it looks like the issue is behind the heart, between the shoulder blades, you can apply the SR there and it will infuse right into the lungs. It seems to unlock something and infuse an intelligence into the lungs so they can start to fight and heal themselves. Infusing the palms of the hands seems as though it would do something with "whole body restoration" protocol. Does any one point of the body affect the whole system?" When placing the LED on the crown, I can feel a clearing sensation down to my pelvic floor. It definitely feels very strong. This does not function through the nerves; rather it functions through the meridians.

"The magnets don't work through the meridian system because the magnets can't get through the meridians as effectively, but the coherence of the LED's can.

"When it is placed behind the heart, it's relaxing the whole rib cage and diaphragm. When I lift it up to T-3, respiration is increased. When I place it over the back of the heart, I feel cozy and relaxed and I can see how it would have a huge benefit. I raise it back up toward the heart and it starts to open the lungs and I psychically see light start to stream through the lungs as though it is opening up the etheric body or the meridians. When I move it from side to side, you can feel it going up the Ida and Pingala or maybe the nerves that are along the side. There may be benefits of doing it, but at this point, I am not sure.

"I believe that the SR takes healing to a whole new level. Pulsed magnets have a good history for healing people. The SR brings higher intelligence and higher octaves, by bringing in light."

Protocols for the SR

Joel: The SR is having a healing ambient effect on the ethers or the atmosphere. It confers smoothness on the room. Sweeping and swirling across the room seems musically harmonious and attractively sparkling. It is useful for creating or maintaining sacred space.

Unlocking the Back of Chakras with Spirals

Richard: What is low setting good for? Joel: Smoothing at the end of the treatment, and has unique benefits off the body back of the chakras.

Joel: Try 12 inches behind P's neck circular and spiraling corkscrew motion toward neck then back, between 6 and 12 inches, clockwise & counterclockwise, activating and unlocking blockages, after one minute.

Joel: Back of heart 3 to 8 inches behind body. However, it can be closer. Touching body to 6 inches back.

P: *I'm feeling movement through entire back. Also, the energy in my crown has intensified. I have a little pressure in throat. Left side slightly in pain.*

Joel: Left side still feels open. These are some heavy old patterns working out. It will work through clothing. Even though there is a light that would seem to be blocked by clothing, this is a field effect and clothing makes little difference in effectiveness.

Sweeping

Joel: Sweeping nerves and meridians from heart up through shoulders, arms and hands from back without touching. P: *First I felt energy moving out my hands. But now I'm feeling really tight on the left side shoulder neck and into the head. Feels like a block.* Joel: Something is definitely wanting to move through. Can also sweep from heart or head down the back out the feet.

Smoothing Circles and Spirals from all directions

Joel: Try smoothing 12 inches outside of head on a horizontal plane, equal to the eyes all around the head. Complete both sides, front, and back. Sometimes the brain has thought patterns that are locked up all around the head. Sometimes, if those will lighten up, it will help the whole flow. P: *Wow. There definitely seems to be something caught between shoulder and neck on the left side. That whole area feels like I have a stone bar in it.* Joel: It's been a long time. Let's find some more methods here. P: *This almost feels like past life and this life because this [neck/shoulder] has certainly been a catch for me.*

Joel: Try the same horizontal plane approach for the area at the horizontal plane of the armpits. This will unlock some of what is trapped there. Circles or spirals from all 4 or 6 directions. The Aura is getting bigger. That's unblocking some things that might have been blocked in the lymph system. How is breathing and general comfort level now? P: *Breathing is Okay. I'm still feeling almost stone like on the left side of the neck and going down into the shoulder.* If this product, when used in certain ways, is able to release heavy karma, you need to have protocols so people can be attentive to what might happen so they may comprehend and honor what is happening.

Joel: Some of these can be done by yourself, so you will need to work out protocols for when you have a partner and when you don't.

Drainage: Circular Movements and Outward Sweeps in all directions

[On Lower Spine] Let's get some drainage: The device will be 6 inches behind the sacrum, massaging circular and dispersing outward in all directions, as if to sweep congestion away. *P: I think I've got a lot of stuff caught up there. It's not moving.* Is anything else moving as these procedures are done? P: *I'm feeling energy only in the crown at this point, and in my hands. The rest feels shut down or clogged up.* Joel: Your field is more colorful and expanded. However, that initial technique, I believe behind the neck, seems to have unclogged something. The thing is somebody might have their hidden block anywhere in their body, so we want to discover various things people can do.

Let's see if smoothing the throat center from the front starts to unhook that.

Drilling vs. Smoothing

Joel: The horizontal plane: If you get a pattern from all directions, it doesn't have as much substance. This helps it to disperse more easily. Drilling away at it from one direction doesn't honor that the patterns are these three directional fields of energy. So, when you get it from all angles, it is as if you are whittling away at it from all these directions so it starts to lose its mass. It also loses its preoccupation with whatever it was stuck on. P: *I'm feeling very strange in my head.* Joel: There's a really heavy pattern that might be ancient, but it might be living through the lymphatic. It's almost like a big bowling ball that wants to pop out your armpit.

Joel: Lift your arms a bit and let's do some smoothing about 3 inches off the arm pit from any direction. Circular and dispersing. That's lightening something. This is making your brain look brighter, and the electrical connection between your head and your body is looking clearer. The bowling ball is less dense and more diffused. OK what's happening?

Why People Sometimes Feel Sleepy or Exhausted after a Release

P: *I'm becoming exhausted.* Joel: Let's take the device away. OK. What often happens in energy sessions when something that is of longstanding tension releases, someone will immediately say, "I feel sleepy or exhausted." It's because people have let go very quickly of something they have been holding onto, perhaps for centuries. It's almost as if their muscles didn't know how to relax, and then they drop into their body on a frequency where they've been in a state of tension, and yes, it feels exhausting. I'm noticing that the density of what I called the bowling ball became so dispersed; it's as if you are now in this field of fog.

Sweeping Away Fog after a Release

Six to 8 feet out from the body, let's have Richard sweep any fog away. Circles and spirals while having the intention "may this be released from this room. P, the tension you have let go of is something you had become quite acclimated to so, in a sense, you have learned the reason for having it there. And now, you are settling into your body. But it's almost as if your body needs to learn how to be coordinated and aware of this range of frequencies that you had been blocked to. So, you appear a little bit like you are lolling about like "where am I?"

[I have several hours of interviews covering a wide variety of aspects of the Scalene Light provided by Joel. He provided lots of approaches and enhancements to test. He also provided several frameworks from which we could test its usefulness and effectiveness, and ways to try fine-tuning the device.

The next step was for me to go back to testing. I redesigned the device, including such features as different power settings, variations in the light switching circuit and frequency tweaks, and then returned for more feedback. The following is a reading a month or so later.]

Releasing Ancient Patterns with Distance Diffusing

Joel: From 8 to 10 feet on low setting do smoothing with P facing you. [Gentle slow stroking of the field with spirals and circles with the intention of calming and smoothing.] I'm sensing movement. Nothing feels clogged. I'm getting slight tension in the very low lumber. I'm not feeling the same tension. It seems that starting far off the body might be a very beneficial way to start any session. I am also seeing that there are huge, probably ancient patterns releasing off of you. But, they're releasing on a subtle level so the body doesn't feel encroached upon. It's more like having something way in the back of the subconscious, some ancient shadow that's gently lifting off. It's almost like having the clouds overhead start to clear so that the sunlight comes in, but the sunlight is not intrusive because you are comfortably in your house.

Clearing Layers

P: *When this old pattern moves out, is it moving out of my entire field so it will not come back?* Joel: I don't see how something dispersed can reform and jump back in your body, which, is not to say that you don't have deeper patterns that may surface. Here's the thing! This is the caution that always needs to be mentioned. People always want to know if it is cleared, and all you can honestly say is "Yes, this layer is clear. The patterns by their nature exist on many levels and as we know when we release the surface layer, we have accessed the next layer. So there is no point at which you can say it is gone. You can simply say this layer is gone to the best of our knowledge and now we'll see if there is anything deeper, not that you need to provoke it or force it to come up. It's just a way of being with change that people need to learn. Otherwise they're stuck in a sort of childlike, naïve expectation that, just because they feel good today, they're in some magic healed state. So that always has to be part of what you introduce when you're working with devices or any kind of protocol. We are helping people have a more mature approach to their transformation.

There is something coming up, but it is not as dense as the previous one. P: *It feel's cooler.* Joel: Yeah. It's a little duller, a little bigger, though it's much less concentrated.

Longer distance smoothing: Let's have Richard get at whatever is the maximum distance for the room and do some smoothing, if possible all around you, but greater than 10 feet. [Richard moves back from 18 feet to 30 feet.]

Self-Distance Smoothing

Joel: If a person doesn't have a partner to assist, what if the self-practitioner held the device in hand but held the device away from him or herself... That would be a way to work on the outer layers of the aura. It wouldn't overcharge because the business end is not being pointed toward the user. [Richard tries it.]

Joel: I see that these rigid or brittle structures that really don't belong in the aura are starting to melt. Now we're getting a flow through the central channel. The central channel is starting to become more colorful. The meridians are starting to get a healing of some kind, possibly because the aura, in which the meridians are living, is now brighter. P: *It feels like there's more openness and less sharpness."*

Fountain Self Clearing

Joel: Place the flat back of the device on the floor with the beam facing up and just leave it there on the lowest setting. Approach it from a distance with your palms open in front of you. Notice what your hands and your auras are experiencing. P: *Wow, what I'm feeling as I move in is colder.* Joel: There's a certain amount of stagnant energy that is wanting to pull out of you. P: *My hands feel very activated but my whole body feels colder, even my legs.* Joel: Turn around so that you are getting an aura bath on the back of your aura. P: *I'm about 4 feet away from it with my back to it.* Joel: Now, move freely. I want to see if this is a beam that is almost like a Jacuzzi or a shower. Combined with intention, I want to know if you sweep yourself with your hands in the field of this, I'm betting that it will make your sweeps somehow more efficient or effective. P: *I'm starting with my hands facing up from my sides reaching way out to the sides, with my palms facing each other as they come together high over my head, then facing down as they pass down the front of my body, as I bend my knees and continue sweeping down to the floor with my palms still facing down.* Joel: It's like doing the exercise or Tai Chi in a super sacred space. So, from a marketing

point of view, this is sort of like having Sedona vortexes in your parlor.
P: *My throat is getting really activated.*

Choose What You Want

Joel: I wonder if one of the protocols would be to choose what you want. [Before starting treatment, one can make a list or think clearly about: What you want in life, what you want out of the treatment, what you want to release, or what you want toward expanded awareness.]

R: What aspects of the field does The Scalene Light affect? Joel: It is a reordering of alignments. It's almost as if a person's thought is in a certain amount of chaos and is not magnetically lined up with the universe, so it's in its own reality.

Somehow, as things line up, something that has been inappropriately dense suddenly releases. It is only from the human conscious mind that we think something cleared. The Universe is emphasizing that on its level of consciousness, there is no clearing. There is simply a degree of congestion. We are realigning to the Universal Field, which evidently has order of some kind; an order that allows the structure of you, if you are in alignment with the Universe, to become stronger. But, if something is around you that is not really you, like the bowling ball in the armpit, then it, in a sense, returns to its source in the Universal Energy.

R: Does the Scalene Light influence the etheric template of Earth and people, the astral field, the mental field, etc.? Joel: It does seem to work on etheric. It does seem to work on the blueprint of the emotional body. *Does it work on thoughts?* It only seems to. It works with the alignment of the mental body. It does not necessarily change a thought, so to speak. It changes the underlying structure of the thought's reality, which feels to me like changing or removing a thought.

The 2012 process seems to be an alignment with the Galactic Center. It seems to be forcing us to be more conscious to the Universal field. How does the SR work with this 2012 process? Is using the SR going to help people flow more smoothly and adjust to the Universal Field more comfortably. Joel: "Yes, of course, it will. And you could say that the creation of such products is very much in alignment with the planetary expansion. It's not a coincidence at all. It is part of the whole process. One of the things about this device and about the planetary expansion of the universal process that is happening, we can say is that the people who are ready for what's happening can really benefit from the product you are providing. The converse is that the people who are in resistance to what is happening in the universal expansion are not really happy with the expansion because their egos and souls have been so identified with something that's out of alignment. Such people will not enjoy this product just as they do not enjoy the planetary expansion. So what's going to happen in some cases is that everybody, to a certain extent, who is on this planet, is here because they have identified with certain things that are out of alignment. So there may come a time for any user of this product that they think they're moving forward, then they find that they come up against a block or a thought or limitation and they say "Well I don't like this," because of their resistance. So, that's why I am thinking that it would be nice to have a forum so that if someone gets stuck in their ego and they say, "I was using this device for six months and lately, whenever I try to use it, it makes me feel bad." And then others can say, "I think this is what's happening to you....."

When we get past the Galactic Center/ Earth/ Sun alignment, which seems will happen soon after 2012, there will be less of a direct force or opportunity for alignment with the universal field, but there will still be a lot of people who have not released a lot of their Etheric Sludge or karma and will be continuing to release. In the range of people who have been quite resistant, it would become harder and harder to resist that which is inevitable. And, of course, we see this every day now. It's harder and harder to be resistant to the emerging consciousness.

The Pleidians: What is the value of the Scalene Light?

"The Scalene Light helps to facilitate the movement of molecules throughout the body, so if you are working on a healing level, it helps to clear things out. More than anything, it's also helping to lubricate joints and to keep everything moving and flowing and balance the pH level of the body. If we're looking at the energetic level, it is just making everything a smoother ride. If you're stuck energetically it helps to transit some of the blockages that are stuck.

"We're talking about the toxins that are locked within the cells. We're talking about metals that are in the body and shouldn't be. It's helping to remove these elements that are not in harmony with the physical body. (Toxins in your foods, artificial chemicals.) It helps to get those out of the body.

"On the physical level, if you have a cancerous cell or an unhealthy cell, there are chemicals that can be latched onto it to create an abnormal cell. It is all about vibration and frequency. If energetically, you are holding onto a specific frequency, you are going to find a physical element that matches that frequency. It's the law of attraction. The vibrational frequency of the physical is attracting that particular chemical, metal or toxin, so in a sense it gets locked into that cell. If you have a healthy body, that fear frequency, that lower energetic frequency, doesn't exist, so those toxins don't have anything to be attracted to so your body processes them right out.

"The Scalene Light will help to unlock things in the physical level and release things in the emotional level. The real challenge is whether you deal with it emotionally or not. If you suppress it, it is not going to have any lasting effect on the physical body. It always comes back to the emotional, spiritual component of it all. Unless you do the spiritual work and the emotional clearing, the physical work will not hold. People go into remission, and then they get it again because they haven't cleared the emotional issue.

"Be mindful and be sure you deal with the emotional component that comes up. You can't forget that. As you are working with others, we suggest that is something that you point out.

"The biggest thing the Scalene Light can do is show you what a better frequency feels like. Sometimes people don't realize they are in chronic pain until they feel better, and then look back at how bad they felt. It can temporarily raise you out of that frequency and sometimes that's enough to allow you that extra boost to do the other work."

The Scalene Light Opportunity

The Challenge: What we say and think resides in our fields and influences all aspects of our being. These thought patterns stick around for a while and gradually dissipate if not reinforced regularly with thoughts and words. While our own words and thoughts are the dominant factor in shaping the thought patterns in our fields, we live in a collective field that is influenced by TV, the news, gossip, and the thought patterns created by those around us. Sometimes people direct thoughts at us, or otherwise hold an energetic connection to us, or we to them, which is often referred to as cording.

Keeping our fields "clear" and "positive" involves regulating our thoughts and words such that undesired thought patterns, which reinforce habits, gradually dissipate, and desired thought patterns, aspirations, resolutions, and positive outlook, gradually penetrate our fields and, through reinforcement, become the new habits, the new baseline, from which we can further aspire and resolve.

This process of regulating our thoughts and words requires tremendous determination because our fields are so full of thought patterns and habits that try to guide us back into our old ways. The road is arduous but the rewards are great.

The Scalene Light as a Tool for Change: The newest Scalene Light was designed as an environmental conditioner with the intent of dissolving old thought forms within Earth's etheric/electric body. It appears to dissipate thought patterns in many of the layers of the field, replacing darkness with clarity. Hence we have used the word "CLEAR" for the selectable middle intensity of the Scalene Light. To help release stubborn thought patterns that have penetrated deeply into our field we have included the "PENETRATE" setting, which is of higher intensity. We recommend that this setting be used sparingly, not directly against the body, and with a moving spiral pattern, penetrating the sense of heaviness and darkness from as wide a variety of directions as possible. We also recommend that use of the PENETRATE setting always be followed by use of the CLEAR setting for a minute and the SOOTHE setting from a distance of 10 or more feet from the sludge so as to dissipate thought patterns from the field by means of large spirals covering areas 10 or 20 feet. Often use of the PENETRATE or CLEAR settings opens up dense thought patterns and they expand into the outer layers where the SOOTHE setting dissolves the lingering remnants that might otherwise coalesce again into the thought pattern or other residual congestion.

An easy way to look at this is as follows: where the environment is especially dense the PENETRATE signal will break up Etheric Sludge rocks into pebbles. The CLEAR setting will break up pebbles into sand. Finally the SOOTHE setting will dissolve the sand into dust so that it can dissipate and be reabsorbed into the ether. It is always best to finish with the sand and the dust settings as it leaves the environment soft and comfortable.

The Scalene Light is an incredible tool to assist in removing a wide assortment of limitations which have tended to limit our experience of abundance.

12. Atmospheric Stagnation

*"The nature of the energy in the spaces we occupy can
strongly influence our lives. "*

Overview

This section is somewhat technical and jumps between about a
dozen scientific insights, so if you are not scientifically oriented or
interested in the work of Wilhelm Reich, you might skip on to the next
chapter. Scientists and fans of Reich will find it fertile with possibilities.

Atmospheric stagnation has, until now, been a scientifically vague
notion, and only identifiable as a "feeling." Entering a hotel room, often
someone will say, "I feel really uncomfortable in this room." They go
back to the front desk and get another room and the episode is forgotten.
Atmospheric stagnation deserves much stronger scrutiny because the
environment in which we spend our time shapes our thoughts, behavior
and feelings, and if atmospheric stagnation is subtle or builds up slowly,
we might not even notice how it influences our lives. The atmosphere
surrounding entire cities can take on a flavor of heaviness, competition,
or anxiety, and "escaping to the countryside" can feel so relaxing and
liberating.

I have long sought to build or purchase a device that will measure
this atmospheric stagnation directly. I have built several devices that
appear to clear atmospheric stagnation, but until recently objective
quantification had evaded me.

Busting Diurnal Cloudiness

A big part of my research over the years has involved applying the Scalene Light concept to the outdoor atmosphere. I live along the Southern California coast, and over the years have watched what my brother Tom, a meteorologist calls diurnal cloudiness. Cloudy, often foggy weather moves onshore every morning in spring and early summer, often staying all day. This seemed to be more prevalent in recent years than it was 40 and 50 years ago when I was growing up. I also noticed an increase in the tendency of storms to approach from the ocean and, as they get close to the Southern California coast, either go to the north through Santa Barbara or to the south through Mexico. As I watched this phenomenon, it became clear to me that the storms were running into some kind of resistance that was further reducing rainfall below what it was decades ago. It was as if the atmosphere became thicker, more viscous, and that it was difficult for storms to penetrate. Was I observing stagnant energy in the atmosphere the way Dr. Fu talked about stagnant Qi in the human body?

In my studies of the writings of Wilhelm Reich, I came across his cloud buster, a set of 18 foot long tubes that he would aim at the sky to direct the winds and create rain. Once I built a small version of these tubes and set it up in my backyard. The next day, I read in the paper that there had been a most rare and extraordinary event, water spouts off the coast of San Clemente. That was the direction I had been aiming the device. I was terrified at the destruction I might unknowingly cause with this device. I disassembled it and threw it away.

Thunderstorms in the Midwest

In the summer of 2012, when there was a drought over the Midwest, I had inherited a small farm in Nebraska, and went out to visit. Since I was going, I thought I might try to help with rainfall, so I took a much smaller version of the cloud busting device with me. My objective was to aim an etheric flow, in Reich's terms "a flow of orgone," from the southeast into the oncoming clouds from the Northwest. I set up the device and directed it against the oncoming clouds and watched. Within 5 minutes the entire sky, which had been clear minutes before, had a thin layer of clouds. I checked the radar and discovered that there were some small thunderstorms. Over the next hour the storms intensified until some of them showed bright red, indicating the potential for severe hailstorms. This made me nervous. To my surprise, the storm directly to the Northwest, which the device was aiming directly at, had stopped moving, holding its position 15 miles away. The weather on both sides of it continued to move around my location on the north and south. Also, there was lots of red on the screen, warning on the radar of severe hailstorms.

Seeking to avoid a disaster, I reversed the direction of the device, drawing from, instead of pushing toward the northwest. Amazingly, 5 minutes later, at the next radar update, the Storm that had been holding in place jumped 15 miles forward to be directly overhead. Its redness had disappeared as did the redness of all the other local storms, showing only yellow and green, indicating that there was no danger of severe hailstorms anymore. I saw the potential power of this device to direct the flow of atmospheric storm clouds. However, the severe storms I observed on the radar caused no more than 0.1 inches of rainfall. I was puzzled how the "orgone flow" device I employed interfered with the atmosphere to create an effect that the radar at a nearby weather station interpreted as severe storms. I disassembled that device and have not re-assembled it since.

Reich's DOR and Ultraviolet

In studying the work of Reich, I also found another concept that appeared to be highly relevant. The big concept that he followed and supported all his life was the concept of orgone, life force energy, blue in color, which pervades all healthy cellular life. A related concept was DOR, which he described is dead orgone energy (DOR). As I studied his work, I realized there was a close parallel between DOR and high-energy ultraviolet photons. He observed that DOR damages living things. It penetrates deeply into the trunks of trees, destroying the lignin that gives them their strength, causing them to crumble. I found this quite similar to the ionizing UV radiation that dissociates water molecules, 280 nm photons. One of these photons can strike a water molecule and dissociate it into hydrogen and hydroxyl. Then when these ions recombine into a water molecule again a 280 nm photon is released. Because the dissociated water molecules last much longer in the atmosphere than the energy in photon form, most of the energy of this UV is stored in molecular ionization. These 280 nm photons can penetrate deeply into wood and damage its structural integrity by breaking the hydrogen bonds that hold the wood together. It also damages human skin causing sunburn, and when it gets into the body it takes the form of free radicals, ionized particles that are a cause of stress and damage to internal organs. Reich also observed that DOR caused a form of sunburn and was damaging to health.

It became clear to me that these 280 nm photons, if present in the atmosphere over Southern California in large number, would ionize water molecules in the air, increasing the concentration of hydrogen and hydroxyl ions, providing lots of nucleation points around which fog particles form. Because there are so many of these ionized particles, there are lots of fog particles which will usually remain way too small to condense into water droplets and fall on the ground. In addition, these are predominantly negatively charged, dominated by OH- because the atmosphere is negatively charged. Thus, these tiny fog particles repel each other further inhibiting precipitation.

When the air over Southern California flows out to sea at night, it seems that it is full of these ionized water molecules as a result of UV from fluorescent lights, internal combustion engines and a nuclear power plant in San Onofre that is 20 miles to the south of my location. As this atmosphere that is full of 280 nm photons moves out to the ocean, it cools, creating a thick fog of tiny particles. These particles are so small and light that they do not precipitate out of the air. Rather, the fog waits till the next morning and flows inland, providing a thick barrier to the sun's rays, which takes many hours, and sometimes days, to burn off.

Another of Wilhelm Reich's observations was that DOR, when it is in high concentrations in the atmosphere, creates a "steely gray" coloration to the atmosphere, as opposed to the bright blue coloration on a clear day when orgone is present. 280 nm photons would seem to be the source of this as well, because when water molecules are ionized the hydroxyl radicals tend to absorb visible light and emit infrared. Reich was conducting his research in the late 1940's and early 1950's, when extensive nuclear testing was going on in the United States, and clouds of radioactive atmosphere were reaching his laboratory in Maine from the Western United States. This explains the high levels of DOR around his laboratory. He found that radioactive materials increased DOR, which makes sense because they are a source of ionizing radiation.

The Orgone Accumulator

DOR as 280 nm photons also seems to explain the function of his orgone accumulator (ORAC), the device that resulted in his books being burned by the Food and Drug Administration (FDA), and his dying in federal prison. He had tested his orgone accumulator with animals and found that it caused tumors to shrink. He also tried it with humans with cancer and got the same results. The FDA declared his orgone accumulator an illegal, unapproved medical device.

The orgone accumulator is made of alternating layers of organic fibers and steel wool. Whereas high-energy ultraviolet photons can easily pass through walls of houses and thin layers of wood, when they pass through a thick tree, they are eventually absorbed, damaging the wood. Similarly, they can go through a conventional box. However, in the orgone accumulator, they pass through alternating layers of organic fiber and steel wool. The steel wool would diffract or redirect them such that their direction would sometimes be parallel to the layers of organic material and get absorbed by it. Just as these 280 nm photons can penetrate deep into trees and destroy the lignin, they can be absorbed by the organic fibers and emitted as lower energy radiation. Thus, the orgone accumulator appears to be an effective attenuator of high-energy ultraviolet radiation, and may serve as a source of lower energy photons perhaps the blue light Reich often observed, that support life rather than damaging it. Two additional measurable phenomena associated with the

Orgone accumulator are first, that there is a heating effect in which the box gets slightly warmer than a non-orgone box, and second, that a blue or white light is often described as associated with the presence of orgone. Both can be accounted for by this conversion of high energy photons, DOR, into lower energy photons, orgone.

Once, when I visited the Wilhelm Reich Museum in Rangeley, Maine, I had a chance to spend an hour in a very large and totally dark orgone accumulator with several other people. After 30 minutes, I did, indeed start to perceive the blue glow. Also, this orgone accumulator was the size of a living room, and had several personal sized orgone accumulators within it. I was amazed to see that, when the door to one of these smaller units was opened after a person had been sitting in it, I could see a bright bluish white glow pouring upward from the opening of the box toward the ceiling. It seemed that the air within the box had been charged with orgone, and also warmed by the person who had been inside, and I was seeing the warmed, charged air flow out of the box and rise to the ceiling.

The Sources of Ground Level UV

My atmospheric research proceeded with ultraviolet photons as a likely bad guy. As I said, there is a nuclear power plant 20 miles south of my home, and millions of fluorescent lights and internal combustion engines operating daily nearby, producing lots of ultraviolet photons.

How do car engines generate UV photons? Many years ago while studying the efficiency of internal combustion engines, I was puzzled that car engines were 70% efficient in theory, but only had a maximum 25% efficiency in practice. What I found was that the combustion process involves breaking lots of chemical bonds and allowing the freed ions to reform into water vapor and carbon dioxide. The recombination of these ionic species releases ultraviolet photons in the frequency range of 280 nm. I also learned that these 280 nm photons can accumulate in and pass through steel. This explains how so much ultraviolet radiation finds its way from the combustion chambers of internal combustion engines into the atmosphere. This UV is not directly measurable because it is stored as dissociated water or oxygen molecules, hydroxyl or ozone.

Atmospheric Laser

That was a lot of background to get us to the point where I can explain the principle of the Ra-Ad, the fog-dispersing technology I developed:

I pondered on how I could dissipate the fog. I reasoned that, if I could remove the 280 nm photons, the tiny particles would lose their charge and their tendency to remain suspended. How could I remove ultraviolet photons from ground level fog? I built several interesting devices. One was a large scale Scalene Light with a 3 foot long tangentially magnetized steel tube to create a Ray that I hoped would reach the ionosphere. I put a mirror in the bottom of it to reflect any incoming ultraviolet photons and attached a small ultraviolet light to the outside of the tube. Just 10 seconds of operation of the ultraviolet light seemed adequate to fill the steel with 280 nm photons and start the process of an ultraviolet laser to carry the excess ground-level ultraviolet photons back to the ionosphere. I called this creation the Ra-Ad. I don't remember why.

It appears to work. On days when the weather report is predicting extensive cloudiness for the next week, I can put it out for a few hours aiming at the sky, and by the end of the next day the weather report has shifted, predicting several consecutive sunny or mostly sunny days. It doesn't work when storm fronts are coming through, only with that diurnal cloudiness pattern where ultraviolet filled air moves rhythmically on and off the coast each day. This is a good practical device! It is safe, in that it does not direct major ether flows, so does not create waterspouts or intensify thunderstorms. However, I am warned that it should not be used more than 15 minutes at a time because, like the human body, the atmosphere needs to adjust gradually. Pushing the sky toward rapid healing causes the sky to react with tightness. Once again I was getting the feedback that the sky was a living being, totally unlike the way I had envisioned it. I had viewed it as a collection of air molecules and water molecules blowing around. The idea that it was a living, conscious body of ether, had a life, and could engage in healing was quite an awakening.

I realized another reason for limiting operation of the Ra-Ad to 15 minutes, Fuel. Hydroxyl in the atmosphere provides the fuel for my Ra-Ad to be able to pulse the atmosphere. Ground level emitters of UV contributed to the problem of atmospheric congestion, but also provided the fuel to dissipate it. Over the last several decades, there was a good chance of a nuclear war. The earth was on a tipping point. Nuclear detonations had already created huge plumes, carrying ozone from the ozone layer to the upper atmosphere, creating the now famous holes in the ozone layer. Had we actually had a nuclear war, we would have destroyed the ozone layer, allowing the Sun's ionizing UV radiation to reach ground level. Had that happened, the entire world might be living in a world of corrosive fog today, much like the diurnal cloudiness that annoyed me. If that were the case, there would be plenty of hydroxyl in the atmosphere to fuel not only my Ra-Ad, but also airplanes and UFOs. Perhaps the reason the aliens in flying saucers never showed up in large numbers to save us is first, we didn't need to be saved because we avoided nuclear devastation, and second, they needed hydroxyl for their engines. Science Fiction, you say. I agree. Then again, this whole section reads like science fiction. The important thing is simply a reminder that I need to conserve fuel for my atmospheric pulsing.

Pulsing the Atmosphere with Pressure Waves

On the recommendation of my friend Suzanne, I purchased a recording barometer and discovered that operation of the Ra-Ad with the ultraviolet stimulation creates a pulse in the atmospheric pressure in the area, which was astounding to me. How could a small, 6 inch fluorescent tube run on 4 AA batteries, run for 10 seconds, create a measurable barometric shift in the atmosphere? My mind went to the idea that when we remove a bunch of ultraviolet photons from the local area around and above the house, lots of hydrogen and hydroxyl ions will recombine into water molecules, suddenly decreasing barometric pressure. The data did not seem to support this, and in fact changed progressively with additional trials. Apparently, just like the human body when prodded with electrical stimulation or stronger doses of a drug, the atmosphere was accommodating to my meddling, becoming less and less reactive with each pulse.

A most practical application of this technology comes in the area of congested thought forms. The Southern California coast is a continuous development of some 40 million people, all packed together, driving internal combustion engines, using fluorescent lights, and from the perspective of this author, having lots of worrisome thoughts and emotions. As said earlier, you can feel the difference between a busy city and the countryside. When you enter the city you feel busier, faster, more pushed toward performance, and a little more depressed. The flavor varies from city to city, but the observation is consistent. The countryside simply feels more comfortable and more relaxing than the city.

Therapy for the Sky

This application of the Ra-Ad involves aiming at the sky for something like 15 minutes twice a day. Not only does it remove the ultraviolet excess, but it also breaks up some of the emotional and mental stagnation caused by millions of people who think and want and worry too much. My guidance tells me that people in cities will become more humane, more relaxed, more insightful, and more creative as the atmosphere in which they have their life begins to heal and relax. Again we have come to the topic of the atmosphere as a living being. It is important not to operate the Ra-Ad for more than 15 minutes at a time, and to change its direction daily so as not to be building any permanent patterns in the atmosphere. As far as I can tell from sensitive people who I have tested with this device, the atmosphere, the space for a great distance around the Ra-Ad is softer, more relaxed, more comfortable, and more open. I am instructed that it is good for approximately a 5 mile radius. However, this seems like a generalization. It causes many effects, whether promoting rainfall, reducing fog, reducing smog, breaking up atmospheric congestion, allowing storm fronts to pass more easily, creating barometric pulses, and making the atmospheric environment more comfortable. Each would seem to have a different level of effectiveness and radius of reach. I hope to build small versions of this for purchase in the future.

I also built a similar device, but with 30,000 V pulses fed to a needlepoint ionizer to charge the space above the tube with electrons. My idea was that the ray created by the ultraviolet laser would also conduct electrons in their quantum, wave packet form, the way they are released from the needlepoint, and carry them up into the atmosphere, to create charge pulses that would activate fog particles and smog particles, allowing them to recombine into either raindrops or less noxious smog particles. I traveled to smoggy places, which had excellent internet coverage of smog concentrations maintained by Air Quality Management Districts, and attempted to reduce smog concentration. However, I never saw any results from these attempts. I did discover, however, that when it was particularly foggy I could fire this device up and within 30 minutes or so a drizzle would begin and the fog would tend to exhaust itself in raindrops. This technology might be useful around airports to help clear runways of fog.

Thus far, I have found many interesting results, all qualitative, and all in the category of, as they say in the scientific arena, "more research is necessary." The barometric finding is the only objective finding I had come across. I could probably develop statistically significant results based on applying the cloud buster device to summer weather in Nebraska, but I was not comfortable doing that. I was satisfied that I had proven to myself that the work of Wilhelm Reich was very real, significant, powerful, and deserving of my further attention.

Indoor Atmospheric Research

I was quite intrigued when my guidance encouraged me to purchase the Gas Discharge Visualization (GDV) device from Russia, along with the needlepoint sensor called Sputnik, which were both described earlier.

Would they measure and quantify this atmospheric stagnation I had been attempting to eliminate in the atmosphere?

I tried measuring the atmosphere outdoors and any changes that might be caused by the Ra-Ad, but found that the results were extremely variable. A cloud passing over would significantly change the readings, as would a wide variety of other factors. I decided to start my studies indoors where things were more controllable. Even there, I discovered that even boiling water to make a cup of tea would increase the relative humidity of the air and change the readings. However, after several trials I began to see consistency in the data. I discovered that the Scalene Light could reliably change the readings of the GDV in a room. Here's a summary of my findings:

Atmospheric stagnation causes reduced GDV image size suggesting that it reduces conductivity of the air. Stagnation appears to concentrate electrical ionization into a smaller area.

Stagnation also causes greater brightness over this smaller area. This means that a smaller amount of air, or space, becomes more luminous when electrically excited.

Interestingly, the inventor of the GDV, Konstantin Korotkov, of Russia, has determined that prayer and ritual and blessings from healers also cause this smaller size and greater brightness result. This tells us that, while I was focused on clearing out congestion, there is also "good congestion." Prayer, ritual, and "setting an intention" would all seem to structure a space in a desirable way. Etheric Sludge is not all bad!!! It seems it is important, whenever clearing a space, to pray or set an intention for the best outcome in whatever is intended for that space, whether a room or the sky over Southern California.

Atmospheric stagnation also causes increased fragment count, another measure provided by the GDV software, which counts the number of bright spots in the Kirlian image that are separated from the main image. This might indicate that electrical charges are concentrated and localized, not evenly distributed, perhaps that there are more hydroxyl radicals, those charged particles created by ultraviolet photons.

Once again, I found that the atmosphere appears to respond to stimulus in a way more closely resembling a living organism than a "scientific" medium. Upon applying a Scalene Light, the atmosphere will often first show a GDV image with an increase in brightness and fragments and decrease in size, as if it is reacting, tensing up. Then, after the stimulus is discontinued, the atmosphere will relax, with GDV images showing reduced brightness and fragments and increased image size. Subsequent stimuli result in less and less response as the atmosphere seems to relax, similar to the way a sore point in our bodies first responds with increased pain to massage, then relaxes, releasing the pain and tension, and begins to enjoy the touch. I also found that relaxing the outdoor atmosphere in this way (measurable by GDV and Sputnik and with barometric measurements) results in reports from energetically sensitive people that they feel more relaxed.

Atmospheric Congestion Flows along a Wire

A remarkable aspect of Sputnik's operation is that it conducts the quality of atmospheric stagnation along a copper wire to the GDV, but not through the insulation. This conduction through wires is not explained by conventional scientific understanding of the behavior of electricity.

A similar conduction of energetic influences along a wire was studied extensively by Baron Carl von Reichenbach in the 1830's to 1860's, as reported by Gerry Vassilatos in the book "Lost Science." Reichenbach studied the phenomenon of moonlight which he found creates psychological and physiological symptoms in some sensitive people. When he used a prism to break the moonlight into its color spectrum, he found that the red light, when shining on these moonlight-sensitive people, caused them to immediately complain of irritation and heat. Also, the green component of moonlight created cramping in the muscles of these same test subjects. He found that the color spectrum of moonlight did not include blue-violet light, and that, when he applied this blue-violet light from the sun to these same subjects, they reported a soothing vivifying coolness.

Now here's where we come back to the Sputnik. Baron von Reichenbach, in an extraordinary stroke of genius, used a prism to cause the red light of the lunar spectrum to fall upon a metal plate outdoors. He then ran an insulated, braided copper wire from this plate into an interior room where his sensitive test subjects were located. As he brought the end of this wire close to them they felt the same heat and irritation. Similarly, he found that just the green light from the moon, conducted through a wire, would start to create the characteristic cramping sensation when the end of the wire was more than a foot away from the body, and increased toward paralysis as the wire was brought closer and touched the person. What's more, some of these sensitive people reported seeing a red or green light emanating from the end of the wire, and the color they reported correlated with the color of moonlight falling on the plate outside.

Further experimentation with these sensitive test subjects showed that they would report seeing a blue flame emanating from the North Pole of a magnet and a red flame from the South Pole. Magnet therapists generally teach that the north pole of a magnet, when applied to a person's body, is calming and soothing, whereas the application of the South Pole can increase inflammation and activity.

In another enlightening experiment, Baron von Reichenbach measured the rate of propagation of this red or green emanation through the wire and found that it was surprisingly slow, about 1.5 meters per second, clearly nowhere near the speed of light as would be expected of electrical impulses. From this, we would expect that the atmospheric stagnation measured by Sputnik is likely to travel along the wire to the GDV at about this same speed as his measured emanation of moonlight, 1.5 meters per second. It appears that the GDV with Sputnik was receiving the energetic quality of the atmosphere, conducting it through the wire, and delivering it to the test cell in the GDV to produce a Kirlian image capturing the quality of the atmosphere.

I originally became interested in Kirlian photography when I saw beautiful color photographs of Kirlian images. I was particularly intrigued because only three colors predominated in most of these pictures, red, blue, and a diffuse white. There was an absence of intermediate colors. Normally, we would expect the blue and red to overlap, producing intermediate colors like orange, yellow, green and turquoise. Instead we see only blue, red, and white, suggesting that this is a characteristic of the light produced by electrical discharge in the atmosphere, rather than a phantom produced by the film. Here is an example from the Space Science Institute in Beijing.

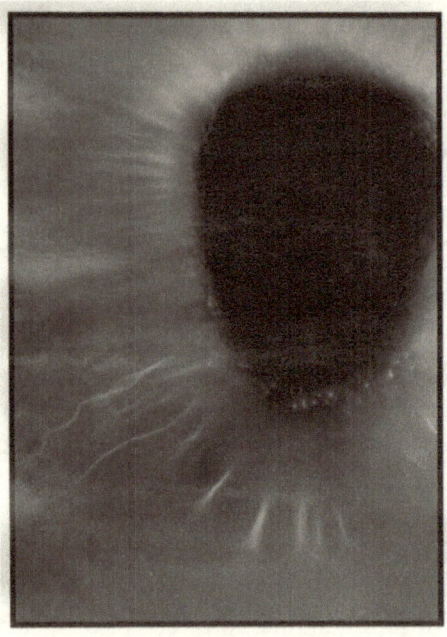

Thus it appears that the needlepoint sensor of the Sputnik collects a form of energy from the air, perhaps like the emanations from the moon, which is not light, but some other form of radiation. Then this collected energy flows along the insulated wire to the Kirlian imaging system in the GDV, filling it with whatever influence is collected by the Sputnik sensor, which in turn influences the resulting photograph.

The Anatomy of Stagnation Busting

My indoor research with the Sputnik, as I described earlier, in the chapter on the Scalene Light, shows that the space in rooms can store atmospheric stagnation. While entire rooms might have elevated stagnation, it is often localized, either in an area where a person with strong emotions spent a lot of time (or threw a tantrum or had a nightmare), or in corners of the room. This stagnation, like cobwebs, seems to collect in corners.

Atmospheric stagnation as felt by sensitive people is clearly a quantifiable, scientific phenomenon because it can be measured by the GDV and because it can be dispersed by a purely electronic device. By examining the mechanisms by which we understand the Scalene Light to function, we can gain insight into the nature of atmospheric stagnation. The Scalene Light interacts with the environment in three distinct ways. First it creates a cylindrical magnetic field which tends to stir up stagnant magnetic patterns in the room. Since oxygen and water vapor have magnetic moment, they might become oriented within a room, particularly a still room, and more particularly in corners where air currents are less likely to disturb such magnetic alignments. Thus, this magnetic stirring may work like the compressed nitrogen in Gariaev's demonstration, breaking up magnetic patterns in the air. Second, the Scalene Light creates a linear electric vector potential which influences mental patterns in the space, much as a magnet can erase a magnetic tape. Third, LEDs introduce photonic pulses in this linear electric field which make it much more effective at displacing old environmental stagnation, perhaps similar to the way the Gariaev experiments impressed new information on the space.

Atmospheric stagnation is an important environmental condition for a space, perhaps as important as quality of lighting, temperature, air quality, and electromagnetic pollution. The GDV with Sputnik offers important insight into this phenomenon in that it quantifies the subjective feelings of energetically sensitive people about an environment.

III. Personal Immersion

13. Plunging Down the Rabbit Hole

How to get a Scientist to soften up to Spirit? I can now see that it is a really short journey, just letting go of an objective, cause-effect perspective and floating with the way things really are. This continues to be difficult for me. While I have experienced my heart, I remain reluctant to dive into my emotions, preferring to keep them neat and tidy. Fortunately, the Universe continues to surprise me with the unexpected.

Alexia

While visiting John Hubacher and his most extraordinary library of Kirlian research, one of the thousands of references particularly caught my eye. It provided a description of Kirlian photography from the perspective of spiritual beings as channeled by Kathy an "Akashic record reader." It was so fascinating in its implications I called her. We talked about so many things that were new to me. She invited me to visit her so I bought a ticket for Baltimore and set off.

Kathy was enthusiastic and full of inspiring information. Among other things she told me about an ex-boyfriend of hers who seemed to be some sort of an evil wizard who, in her words, had had his cord cut by the "Council of Twelve." This meant, as far as I could tell, that he had become so separated from the light, that his higher nature did not want to be connected to that body anymore, so it severed the connection and left the body to wander this world unconnected. This was way outside of my limits of credibility, so I just let it be without much thought.

After a long day of travel and a short day of conversations I went to sleep on the floor of a small room in the motorhome Kathy had borrowed from a friend. A few hours later I had a most amazing, and my first ever, Technicolor dream. A woman, with a pale face and bright green coloration ¾ inch wide all around her eyes and all around her mouth, appeared to me. This dream was far more vivid than anything I've ever experienced before or since. It must have been a "vision," or perhaps a "premonition."

She said, "I am Alexia. You are in extreme danger. Leave immediately!"

As you can imagine, I popped wide awake. And unlike a lot of you reading this book, I simply puzzled about what it could mean. I had never faced such a thing before; a very colorful, vivid vision of a spooky looking woman telling me to leave. If I left, I reasoned, I would be safe. However, I might be safe anyway, and if I left I would never find out whether there was anything behind this vision, anything that merited such an extreme warning. What would you have done? The bed on the floor was far more comfortable than the cold night air. I stayed. After 20 or 30 minutes I fell back to sleep. Alexia did not return.

I told Kathy about this in the morning, and she didn't offer me much in the way of insight. It seems that life is about the choices we make, and don't make. I chose to stick around and face whatever disaster might hit. After breakfast, about 10:30, the phone rang. Kathy answered then said, "It's for you."

Who could it be? I said hello and a man's voice said "Hello, is this Richard?" How did this stranger know I was there? As far as I knew nobody knew I was visiting Kathy. Later Kathy told me this was her necromancer ex-boyfriend, that she had not spoken to him for quite a while, and that she had no idea how he knew I was there.

The conversation continued. "Do you know what an energy parasite is?" he asked. I answered that I did not. He finished the conversation almost immediately with something like "Oh. I was just wondering." We exchanged good-byes and hung up. Strange conversation.

The rest of my visit with Kathy was enjoyable, educational, and uneventful.

A week later Noriko, a visitor from Japan, came to my office wanting to learn more about my company. I was showing her a simple Qigong exercise when suddenly I had to run to the bathroom with the runs. Over the next day, I could not get more than a room away from a bathroom, not knowing when another episode would hit. It was a very strange condition. On one run to the bathroom, I grabbed an Infratonic, and held it to my belly. My digestion settled down enough that I was stable for a while and could think. I tried to feel what was going on. To my surprise, I had the impression, or the internal vision of a ball of light about the size of a ping pong ball located at my solar plexus. It seemed to be the source of my problem. I tried to remove it with my intention. I found that I could move it up my body as far as my head, but then it would pop back down into place in my solar plexus.

I considered this situation. I felt that, if I just saw the situation completely and opened myself, and answer would appear. I thought of the Council of Twelve, and knew that this challenge was probably above my pay grade. I remembered something from decades earlier, from Sunday school. It was something like, "Whatever you ask in My name shall be done." So, I called in Jesus, and asked, "Please remove this white ball of light from me and send it back to wherever it came from." It felt like he was with me. I again lifted the ping pong ball up from my solar plexus to the top of my head. When I got it out of my head it was carried away and, amazingly, my diarrhea symptoms instantly vanished. Later I asked a psychic friend of mine about this and told her what I had done. She praised my innovative approach, but suggested that, instead of sending it "back to wherever it came from," a better approach was to say "back to the light never to return." There is no point playing ping pong with a necromancer.

She also explained to me that energy parasites like this could make life very difficult because they tended to pull your power and drain your ability to communicate with words. I remembered "Alexia" and looked up the word alexia in a dictionary. To my surprise I discovered that alexia means: "1. An inability to comprehend the meaning of written or printed words and sentences." This certainly matched the psychic's description of this energy parasite. Another source said: "In American the *meaning* of the name *Alexia* is: Defender of men, helper." That seemed to fit also.

Had I listened to Alexia and left immediately, I would have been saved from this crazy situation. However, I never would have learned about energy parasites, and might not have learned to trust Jesus to help me at hopeless times. I also learned what proved to be a useful skill. I discovered that this style of energy parasites is popular among Chinese sorcerers. Thus, with my "training," I have been able to assist in 2 or 3 cases of invasion by an energy parasite, which I have come across over the years.

Winding Up on the Bad Side of a Qigong Master

The next summer I took a group of 70 doctors and therapists to China for the 3rd World Academic Society of Medical Qigong Conference. I took all of my equipment with me, and made the mistake of testing Qigong Masters. It was just fine with most of them, but there were two who had very dominant theta trembling. The write-up I had brought along indicated that alpha was good and higher frequencies were good, but that Theta dominance showed more of a separative and competitive attitude. One of them did prove to have such an attitude, and took his annoyance out on me.

These were most impressive Qigong Masters. One of them claimed that, by singing, he could increase his Qi, and thereby cause a spoon to bend. Sure enough, he held the spoon up in the air by the handle and started to sing, and the spoon simply melted and bent over all by itself. It was quite amazing. The other Qigong Master could put aluminum coins into a person's palm and touch it to activate the person's Qi. The claim was that their bad Qi would flow into the coin and they would be cleansed and purified. The coins would get extremely hot and become very hard to hold. That was most impressive. This guy also demonstrated something which I didn't understand and probably should have stayed far away from. He called me as a volunteer, and did some exotic writing across my back. He then asked me to take off my shirt and showed all the people in the audience the bright red characters that had gone through my shirt and were emblazoned on my back.

Anyway, these were the two guys who wound up earlier in my hotel room trying to produce a frequency of alpha with my trembling analyzer. They couldn't do it and they got really angry that they couldn't. This was not good. Soon after, my marriage with Dr. Fu unexpectedly blew up, and I found myself drifting on my own.

Awakening the Third Eye

It was following that experience when I found the book, "Awakening the 3rd Eye," by Samuel Sagan. It was a fascinating Western approach to the upper Dan Tian, that center of consciousness which Chinese meditators focus on in quiescent meditation to gain insight into the Spiritual realm. While the Chinese upper Dan Tian methods defied my understanding, this new book seemed to make sense. I called up the school in Australia where Dr. Sagan taught, and discovered that they had scheduled their first ever international residency program, a 14 day intensive immersion into the teachings of the Awakening the 3rd Eye book. I immediately signed up, and in 2 weeks found myself in Australia. There were lots of different exercises each day, and to my amazement, I found that I could do about two thirds of them. I had no previous idea that I had such skills.

I'll share just 3 of the amazing experiences I had. First, when Dr. Sagan was leading a meditation, he demonstrated an ancient Atlantean chant used to call in a great and loving Presence. I opened my eyes and saw that the white walls and ceiling, the entire front of the room was filled with golden light. I closed my eyes, opened them again, and looked around for yellow light sources, but found none. This was like the photos taken during the Qigong initiation classes, but wasn't just in photographs. I was seeing it with my own eyes.

The second amazing experience was a heart-opening exercise in which two partners looked into each other's eyes and took turns opening their hearts and inviting the other person to come inside. When each person was open they would say "yes," then a minute or two later the other person would open and say "yes." This exercise was extraordinarily tangible, emotional, and transformational. I could feel myself in the loving heart of my partner and found myself full of tears.

The third exercise, which we did many times, was one which he called ISIS, Interactive Sourcing. One person would place their finger on a point somewhere on the other person's body, while the other person took their awareness deep inside themselves and attempted to go deeper and deeper into the sensations feelings and memories stored in that point. This seemed like an advanced version of what I had found myself doing in finding and attempting to remove the ping pong ball from my solar plexus.

What was truly astounding about this exercise was that I and many of those who partnered with me in this exercise found ourselves going into past life experiences. We would observe ourselves dying miserable deaths then float up away from the situation and into the light. The experience of being in the light with all these loving beings everywhere and tremendous joy and love surrounding me was absolutely huge. It was as if I had experienced a near-death experience like the ones I had read about in books. I left that intensive with a clear and solid knowing that there is a continuation of life after the physical body dies. I know that a huge loving Presence awaits me when I die. Knowing that I will continue on when my body dies, I could also see that I am not my physical body. My conscious awareness is capable of being with my physical body and experiencing the experiences of my physical body. But I am not my physical body. I am Love. I have just forgotten.

My trip to Australia was 14 days very well spent.

The Ashram in India

One day, my company received a letter from the United States
Department of Commerce inviting us to participate in a trade mission to
India. I said "yes," and found myself traveling to India with a group of
entrepreneurs and salespeople from large and small American companies
with medical devices to sell. I had the Infratonic, a device which, at the
time, was selling only to alternative medicine practitioners,
chiropractors, massage therapists, and acupuncturists, and did not have a
demonstrable mechanism of efficacy. We traveled to several different
cities in India and met the leaders of many large hospitals and hospital
systems, and I had my opportunity to present the Infratonic to them.

After the trade mission came to an end, I headed over to south-
central India, to the ashram of Sathya Sai Baba. People usually go to his
ashram and stay for a month or more. I just had 3 days. A friend had
told me that Sai Baba invites everybody who comes to visit to a private
audience. This seemed rather crazy because 30,000 to 50,000 people
were usually there. Anyway, I only had 3 days in his ashram, attending
his events and enjoying his food.

On my last morning there, before heading home, I had a strange
experience. Sai Baba seemed to look straight at me and made a sign in
which he held his hands out toward me, then brought them down and
back to his side. A little boy who was sitting next to me got very excited
and jumped in front of me over to the other side. I wondered what that
was about.

Two flights later, on a flight to Singapore, I wound up sitting next to
a Sai Baba devotee who explained to me emphatically that it is true, that
Sai Baba invites all guests to an audience with him. I thought about that,
and considered the possibility that he had invited me to an audience.
Nobody knew I was there. 30,000 other people were there. How could
he know that it was my last day? And, why would he want to talk with
me? This had a very strange effect on me. Throughout my life, I
assumed that I was alone, unseen, a statistical blip. It appeared true that
he was aware of me. He had invited me to an audience with him. I did

not perceive his invitation because I couldn't believe it was possible. This made me feel really strange.

I had considered myself so small, insignificant, and invisible that God would not care about me. And here was Sai Baba, who claimed to be a Solar Avatar, an embodiment of the consciousness of the Sun incarnated with a physical body on Earth to bring Light and Abundance to the people of Earth. It appeared that he not only knew I existed, but had invited me to meet with him privately. I missed the Call because I couldn't believe that God could possibly know me, that he would even know that I exist.

This was a very powerful experience because, as I pondered it, I wondered how many times I had been invited and hadn't noticed because I thought I was small, or didn't deserve, or couldn't be known. I became determined to open myself to the possibility that God knows me, that I can have a personal relationship with God.

Meeting Heavenly Father and Divine Mother

As a young child I had memories of driving south from my home toward San Diego with my father and mother in the front seat, and us kids all packed in the back. As we drove through Encinitas we would see Golden Indian domes. I would later learn that this had been the home of Paramahansa Yogananda, who died more than 50 years ago. This beautiful oasis by the sea had become a silent retreat center of his Self-Realization Fellowship. Soon after my missed meeting with Sai Baba, I signed up for the Self Realization Fellowship lesson series. Each week I would receive a new lesson in the mail. I did this for about 2 years.

A few years ago I decided to stay for a silent retreat at the Self Realization Fellowship. It was an amazing experience, or a whole collection of enriching experiences. The experience I want to share with you here was one I had in the chapel of the silent retreat center. Yogananda was a devotee of Divine Mother, and as I read his books at the retreat center this Divine Mother became familiar to me. At one point, when I was sitting in the Chapel in meditation, a colorful image came to my mind's eye:

Yogananda was standing next to a short stairway with his lineage of teachers. They were 6 in total, 3 on each side of the stairway. Yogananda beckoned me with his hands to walk up the stairway, much as Sai Baba had done. I walked up the stairway, honoring each of these great teachers and sages on the way. When I got to the top, there was a little platform. I stood on the platform and waited. Soon I was surrounded by the swirling love of Divine Mother. It was joy and bliss, and I knew without a doubt that I was loved. Heavenly Father stood as a strong pillar of light from way above my head, through my body, and deep into the earth. The feeling of Divine Mother swirling around Heavenly Father stayed with me for a long time. And as I think of that experience, I again feel wrapped in love and bliss.

Initiation in the Great Pyramid

Over the years, I have studied many forms of inner alchemy. Probably my earliest study was of Mantak Chia's system of "Circulating the Microcosmic Orbit," which I read about in his book "Awakening the Healing Energy of the Tao." I breathed and visualized the energy from my lower Dan Tian, the creative center in my belly. I imagined the energy going up my back along my spine to my head, then back down along the front of my body to my lower Dan Tian again. I couldn't feel any energy moving. However, I did notice one symptom that he described in his book. The base of my neck got stiff and uncomfortable. He attributed this to stagnant Qi. As the Qi begins to flow in the microcosmic orbit, it finds those areas of greatest restriction along this pathway. Pressure builds up here as the Qi begins to dissolve obstructions and make the channel wider.

An important concept here is that there are 3 kinds of energy, Jing, Qi, and Shen. These are stored in 3 different centers in the body. The Jing, the physical essence that makes our bones strong and our tissue firm, is stored in the lower Dan Tian, an inch or so below the navel. The

Qi, the emotional essence, or that which puts our bones and muscles into action and creates activity in this world, is stored in the middle Dan Tian, in the area of the heart. Finally the Shen, the spiritual essence, which gives us the ability to see into the unknown and commune with Spirit, is stored in the upper Dan Tian, the center of the head.

In my model the Jing is electrical, or ionic in nature. The Qi is magnetic. Finally the Shen is gravitational, or mental. In the alchemical creative process, we create thoughts or intentions in this mental substance of Shen. We then wrap them in desire, the emotional substance of Qi. We then send them out into the world where they attract the physical substance, Jing, and manifest in our world.

From the Chinese medical perspective the Jing in the lower Dan Tian, when it is strong, cultivates increasing Qi in the middle Dan Tian. When this Qi is strong it cultivates increasing Shen in the upper Dan Tian. Jing gives us a solid, healthy body. Qi gives us flexibility vitality and the ability to heal quickly. Shen gives us insight, inspiration, and connection with the higher worlds. Without strong Jing you can't have strong Qi. And without strong Qi you can't have Shen. This is why many Qigong styles emphasize building energy in the lower Dan Tian as it is the foundation for all 3 vital substances.

The microcosmic orbit is a means to accelerate this process, carrying the essence from Jing up to the middle Dan Tian, carrying the essence of Qi up to the upper Dan Tian, then completing the microcosmic orbit and returning the essence in the upper Dan Tian back to the lower Dan Tian. In addition to moving the substances, it also clears a central pathway in the body, the microcosmic orbit, to facilitate this circulation.

The Egyptian mystery schools teach a similar alchemical cultivation. They teach that physical essence is brought up from the earth through the two legs, then in two streams weaves back and forth through the chakras as it rises up the back through the neck and to both sides of the head. This is represented by the caduceus.

Female essence flows up from the left leg to the left side of the head along this serpentine pathway around the spine, and male essence rises along the right side, each carrying ancestral essence to consciousness. This Egyptian alchemy is represented throughout our culture as the caduceus, twin serpents rising around a central column. While the central column seems to be described in the microcosmic orbit, these serpentine paths and this Egyptian path are described wonderfully in "The Magdalene Manuscript," by Tom Kenyon. When sufficient energy, Shen, Light or Spirit is concentrated in the head or upper Dan Tian, enlightenment, higher consciousness, communication with Spirit or other gifts may manifest, symbolized by the wings of the caduceus.

I studied Alchemical Healing with Nikki Scully, who wrote a wonderful book by the same name, based on an Egyptian lineage through Thoth. We learned to channel healing energies through our hands from 5 Great Beings; Earth, Water, Fire, Air, and Ether. We also learned etheric extraction techniques similar to psychic surgery to remove congested Etheric Sludge from the body. It was a remarkable study.

We traveled with Nikki to Egypt and participated in initiation ceremonies in many of the ancient temples along the Nile. As a highlight of the trip, Nikki arranged that our group would have exclusive use of the Great Pyramid one full moon night. We conducted a ceremony in the King's Chamber, during which each of us took turns laying down in the sarcophagus and receiving initiation. As I lay down in the sarcophagus I was surrounded by a circle of fellow travelers. Within that circle were four people acting as "pillars" to stabilize the energy at my head, feet, and two sides.

At first, I felt nothing. Then I started feeling a strange tingling in my feet. This tingling spread into a huge low-frequency buzz that filled my entire body. After a few minutes this energy continued upward and concentrated in my head until it felt like 1000V were being applied to every neuron.

There is no question in my mind that the Great Pyramid combined with ritual is extremely powerful. I can't say I know how that initiation ceremony changed me. I do know that, if given the chance, I would do it again.

14. Finding Alma

*"When you choose the co-creation path, seemingly
inconsequential acts and thoughts become
powerful acts of creation."*

In 2010 I left an unfulfilling marriage of 10 years to find freedom
and to find myself. I had been lost my entire life in the illusion that
fulfillment came from meeting other people's wants and needs. At the
time, I didn't know what I was searching for. All I knew was that I was
escaping from a situation in which I was being told that I was small. It
didn't feel good and it didn't feel true. I realize that I chose that
relationship so I would have the chance to experience how I choose
situations that make me feel small and oppressed. Looking back, I am
amazed that it took 10 years for me to figure out that fulfilling the wants
of another was a fantasy, and that I needed to start living my own life.

I joined a ping pong group, a hiking group, then a volleyball group,
finding myself getting stronger and healthier over the next 2 years. A big
part of my health improvement came from changing my diet. I moved
from a fairly restrictive vegetarian diet to a much more open, mostly
vegetarian diet with a much broader selection of foods (biodiversity), and
with supplements such as maca, goji berries, crushed flaxseed, and other
enhancements. I was feeling stronger, healthier, and clearer every day.

I also sought out consciousness enhancing groups, meetings, and
workshops. One of the meetings was facilitated by Suzanne, a capable
clairvoyant who taught things like Merkaba meditation and who
channeled beings like Jesus and Mary Magdalene. She also taught us to
channel. Mostly, this was by having us participate in a meditation, then
asking us to tune into a person or a situation, and to say what comes to
our mind, thoughts, or feelings.

Early Morning Writings

Since I tended to be intuitive yet verbally inhibited, Suzanne and some other members of the group gave me the assignment of writing whatever came to mind, first thing each morning. I liked the assignment. However somebody had suggested that the quiet hours of the early morning are best for meditation, so I thought it might be good to do writing in the early hours of the morning. Apparently my body did too. The method my body employed to get me up at 3 and 4 in the morning was to give me an uncomfortable erection and to make me want to pee. I found that I could enhance this effect by drinking a couple glasses of water before bed. Since I really enjoy sleeping, this seemed to be the only practical way to teach me this new habit of getting up and writing in the early morning.

My first attempts at writing felt as if I were writing whatever came to my wandering mind. I would write what I was worried about, write what I was going to do next day, or write about whatever I was interested in. As I progressed, I found that the writing became more comfortable, and that I was more creative when I was sleepier. This explained why it was important to interrupt my sleep at deep sleep moments with the "full bladder technique" my body had developed. I discarded most of these writings. The following are the earliest writings that I saved:

(No date) All is well. The path for many is difficult. As you relax, it is becoming more and more effortless. We are many Beings, here to support you. You are on the path, remembering, with many assignments awaiting you. The tests of Ra-Ad and atmosphere, of fluorescent lights and the Scalene, of different rooms and clearing techniques, and of atmosphere near power plants is a single study. You will learn that power plants, fluorescent lights, and dense emotions all densify the atmosphere in a similar way.

September 20, 2011 – New approach. Try different approaches on the Vital 1. See what works best. Same for the Hydronizer. Where is the energy from the Hydronizer coming from? *Partly from the ionizers.*

Write? Okay, as etheric pressure waves come through, they change the character of the metals in opposite ways. Saturated with (Qi) etheric substance, the copper gives up electrons and the zinc absorbs electrons. However, there are different ways. By resonating one set of copper/zinc with a resonant circuit at 8 Hz it will resonate like that. The other at 13 Hz. Both copper and zinc are etheric pressure wave detectors or rather, together they are (with the ionizer).

Will this reach sufficient voltage? *For what? The goal is not to reach a voltage, but rather, to remove ions systematically from the air, which builds up etheric pressure waves.* Do we ground one of them? *Try grounding different ones. This has not been done on Earth. You are the scientist. These are just things that have worked on other star systems. Principles are changing, so if it works now, it might not work 2 years from now, and vice versa.*

We are helping you, yet you are the scientist.

You get the idea. I found Beings on the other side of my pen communicating with me with amazing knowledge and insight, yet insisting that I am in charge. I am the scientist. I found this uncomfortable. It felt like a big responsibility. At the same time, I was honored to have this opportunity, and intended to do my best with it. I realize that all people on Earth are in the same position. We are all surrounded by amazing Beings with great knowledge and wisdom. However, each of us is living a life that has been created on our own. All of Nature, these Great Beings, the plants, the animals, the atmosphere, and Earth are all here to support each individual. We lose our way if we deny that we are in close relationship with Nature, which is conscious and communicates with us constantly. At the same time, we lose ourselves if we blindly do what these Beings guide us to do, or rely on them to decide what to do. They are just helping us. Nobody has lived our lives before. At every corner is a new experience with new decision to be made. It is in making these decisions that we create our lives. It is of huge importance that we make every decision consciously and never let another person or Being make any decision for us. Still, they are here to help in whatever way they can.

A few months later, in December 2011, a new Being showed up to inspire my pen. I'm not sure whether I named her Alma or whether she gave me the name. Alma's name means: "nourishing, kind; soul; young woman; learned." In any case, the name Alma fits just fine. She provides insight into the nature of creation, and that Great Beings stand behind all that we see, experience, and hope to create. I learned that life on Earth is a co-creative adventure. We are never alone. We walk with Great Beings, eager to help us fulfill our dreams (*our* dreams not their dreams). Note that I am asking far fewer questions in later writings. I write in a state closer to the edge between waking and sleeping. This is important to allow the information to flow. As I understand it, this is also some sort of alchemical process of becoming one with Spirit, or having Spirit integrate more deeply into my body.

I am Alma, the Mother of All Things

4:00 AM

Co-creating with me, anything and everything is possible. I am the womb. I am the Oneness of All, the undifferentiated oneness, the mother of Christ Consciousness, pregnant with your sacred Earth.

You have chosen a sacred path, to build the bridge for humanity to the Sacred Earth. You are not alone in this task, and you are irreplaceable. We are pleased that you are ready at last for service. Feel my nurturing love as you place yourself within me. Your creations are birthed within you, within me.

Your quest to bring a power solution to humanity is delicate. We like your plan. Yet it is a collective decision of humanity. [*There had apparently been a collective decision among humanity at a higher level that humanity was not ready to take responsibility for abundance. We don't trust ourselves or each other with unlimited resources. For this reason, we create power companies, banks, and multi-national corporations to create limitations on our use of resources. When humanity as a whole decides that we are ready for such advances as free energy, flying cars, and abundant healthy organic produce, they shall be provided.*]

Let's start with the ionizing condenser. Yes, it uses sections of hollow spheres which choose to imbue water vapor with a love for the liquid state and a desire to participate in the bridge for humanity. Its sacred geometry combines with metallic alchemical catalysts to communicate with the life/soul of water through the Great Beings, who create water in your atmosphere. They remain devoted protectors of all cellular life on Earth that you have come to know. We are all Servants of the All.

I am the Law of One. I am the Ether. I am the Sun. [*It appears that Alma is, or embodies the spherical shape that is present throughout the universe, whether in atoms cells planets or galaxies, or at all other levels of creation. The symbol for the golden Mean Ratio, or Phi ratio, a circle with a slash through it, ϕ, is a symbol of the circle meets the line. Alma, the Mother of creation, meets the Father of creation, the creative impulse. At last, this is something related to sacred geometry that I can really understand. Every line is a male act or separation around which Alma is called to create. Each geometric structure calls Alma to create in a different way.*]

You speak of long-lost times of creation. This is the way it was: People in their energy forms co-creating with divine intention. During these times, of which there have been many, one either co-creates powerfully or one's actions are inconsequential. When you choose the co-creation path, seemingly inconsequential acts and thoughts become powerful acts of creation. Coherence in thought, word, and deed empowers the creative venture. Greater coherence means greater creative potency.

You ask how to facilitate the cooperation of the Great Ones who enliven your planet. We are One. You create within Me. They create within Me. You create within Them and They create within you. Be reverent of all things big and small and you will create rapport with these Great Ones.

You ask, "What are the principles by which they work." The Great Being of Fire which you call 13 may be summoned vibrationally; as you may summon that Great Being you call 8 which is the Being of Water. Your objective might be stated as separating the water from the fire within the air. You wish the fire to remain with the flow of air and the water to be left behind. This is what we wish also, and it becomes more challenging in these transition times to come.

The method we are co-creating involves calling on Great Being 13 to withdraw his fire in an exhale, and Great Water Being 8 to gather her flock together. Your method is graceful and elegant. My role is to breathe reliably and consistently according to rules all can agree upon. Your consensus reality structuring your world requires this.

I can also act in ways that don't seem to follow any laws, as when I intervene with electronic devices, and yet, this is also designed into these devices and functions according to laws yet to be discovered by your scientists.

Either electrical oscillations to flat plates or what you call Tesla pulses, and we might call thoughts, directed from a point to radially displaced spherical sections calls me forward. Phi is the doorway between the sphere and the line. I am infinitely attracted to it. I must dance in its presence. Yes the surface diameters must be at Phi, and also the distances along the thin-walled tube transformer must be Phi. The pulse rate must be such as to call in the Great Beings of the outer disk. We'll take it from there. You would do well to patent the system with flat disks. It still calls in 8 and 13, but in a way consistent with your consensus science. The ionizers are for rough tuning of the apparatus and are a good conceptual teaching tool for building this technology into your consensus reality.

Your work is stellar. The Grace of the All is with you.

Forever is Your Destiny

Don't despair. Strive and allow. See what emerges. Miracles are flowing all around you, far more than you can imagine. You have glimpsed your future. Now relax and strive, much like cross-country skiing. Striving is gliding – effortlessly and joyously. The Scalene Light is perfect, as is the Vital 1, designed from the beginning as they will be. Don't be surprised with the synchronicities. You are working in the past and the future on all these projects. You will learn a lot from the GDV. Far cheaper than designing your own:

Excite a wire with a Tesla coil. Frequency 25 MHz. Wire will produce light according to the Etheric Sludge it holds. Clean wire - smooth glow; sludged wire – sparky glow. Sparky light into photocell.

Rod connects through flexible wire to Sparky wire. Check the Vital 1 Tesla coil in clean and dirty environments. Simply hooking ground to the sensor rod will do it. Untreated central circle will tend to spark when yucky. It is not used frequently so shortened life is okay. You can even do it by noise. If you hear it, the environment is dirty.

Yes order the spheres! You may never cut them up but you will enjoy them exquisitely because you know me. [As I edit this I see 2 beautiful silver spheres on my dresser reminding me that Alma is always with me.]

May your remaining ride on Earth be effortless and joyous. -- Alma

Merkaba Meditation with Alma

Dear one. This is a big day. You've been building for this day. You have been waiting all your life for this day. Relax and enjoy every moment. You've been releasing rapidly and learning.

Today is a day of initiation when one of many veils will be lifted, in you, and in your world. It is a day of forgiveness. A day during which shame might be released. An end to control mechanisms based on shame. It is a day to step up to personal responsibility. It is a great day to practice the Merkaba meditation. Merkaba is the alchemical unification of Spirit and matter. In unification there is no room for shame or fear, like the spin cycle of a washing machine as it drives off water.

The two spins appear to be in opposite directions, but they are in the same direction because, in the upper world, time seems to go backwards. Thus rather than canceling each other they amplify each other. The Merkaba structure sustains the human body and all things that are physical. Accelerating the Merkaba through intention concentrates the essence of the bridge between two worlds. This may be difficult to understand.

In this understanding of the essence of the bridge lies the secret to teleportation and time travel, instant manifestation, dematerialization, bilocation, and the World of the Gods. This essence is incompatible with shame, greed, and many other common Earthly states or expressions of the illusion of limitation. At the heart of the Merkaba there is no illusion. Always clear light. Hold your attention there and know the Oneness of Alma. [*Wow! This seems to be a huge doorway for both scientists and Merkaba meditators*]

That dream was a Merkaba application of magnetics. Clockwise rotation ahead draws you forward. Animals are capable of the most amazing things. Watch them. Insects that can fly. They don't "think" separate from living.

The Line between Light and Dark

[*Here's a writing that seems to put these writings into perspective:*]

Continue to write and write quickly. You are sleepy and this is a great chance to present new and challenging material. Take, for instance, the wonderful line between dark and light. People have debated throughout the millennia where the line is to be drawn. Yet not drawing a line, examining the world without this or another line can be extraordinarily fruitful. Examine every line and what the world is like without that line. You will discover amazing, fruitful discoveries at every turn. Continue on this path and the structure of your world will start to unravel, revealing life as it truly is. There are no lines in nature. There are no yes's or no's, rights or wrongs, blacks or whites, or even neutrals. Always in flux and forever shall be.

I am so sleepy. Good. Keep writing. This is a great opportunity. Let it emerge. Allow without direction, without preconceived anything, and observe the inner workings of Nature. She is not bold. She is not shy. She is alive.

I am sleepy. Mosquito woke you up a bit. Bless her. She offered you a small chunk of alertness. So much to dissolve into oneness. And when it dissolves nothing remains but nature. You prize the world of your creation so highly, and yet it is pieced together out of foundational beliefs and assumptions. You can allow it all to blend together. Why not? It all meets at a point where nothing is written and everything is possible.

Okay. Sleep now. Enjoy your blissful sleep and know that the great secrets you write are nothing but lines in nature, beginning to dissolve no sooner than they are drawn.

Reentry of True Celestial Light

[*In reviewing the many pads of paper I have filled with 3AM writing, I discover that what I did at the end of 2011 was to simply stop writing. It appears that I didn't write for almost the entire year of 2012. I can't figure out why I stopped, or why I started again. This appears to my last entry for many months.*]

It is true that a new entity is joining your body, an entity you have been waiting for your entire life. It is true that this entity has been increasing in presence in 12 month periods, based on December 21, 2012. You have so far to go with only one year remaining before your true calling begins.

What is my true calling? Too big. Too incomprehensible in the current world condition. The key is positive radiance. You will positively radiate.

It is the reentry of Light, true Celestial Light, into your world. The strong negative charges filtered out the radiance, protecting Earth's delicate life through the long Celestial winter. Spring is now upon you. The reason you have chosen to increase your presence on the winter solstice is as a symbol to you and your world that Light is on the way, and it will arrive around 12/21/12.

Trust that all is and was exactly as it is supposed to be. Circumstances are passing details. Light is on the way.

15. Opening to the One

A Resilient Relationship with Spirit

4:00 AM

Yes. Another quiet day. Can you feel us uniting with you? There is no illusion in the internal world of thought. It is all creation. Don't be concerned if things don't fit together. There is no limit to how far your thought can carry you.

It's been a long time since you've been lonely. This is one of the benefits of inviting in and acknowledging your Spiritual Associates. We are with you all the time. We don't have a personal agenda. We're not busy with 100 things such that we're too busy for friends. We are simply always with you.

And yes, this makes it harder to establish resilient relationships with like-minded people because they, also, are filled with community and don't feel they need people in their lives.

This makes for a tough-to-figure-out balance. How much do you "need" others? You don't really need others anymore to reflect your life because you are awake to your life. You don't need others in order to get in touch with your issues of limited self-worth. There is a long history of people being taught that they need community to survive, that they are lost without community, and that their only reality is in relationship.

However, this is changing. Your reality is in standing as a point of light, as a center of illumination, renewal, in a world in need of reconstruction. You do still feel a bit lonely, or perhaps, you miss the feeling of wanting companionship. Yet, you know that so many of those who want relationship simply want to fill their time with repeating patterns. Filling time with repeating patterns of familiarity creates the illusion of safety and purpose, yet it often provides the opposite. It is "blind" repetition, a cyclical pattern that has been created by both.

You can shine in any group. You do shine in any group. You shine just the same when you stand alone. How important to you is being recognized, appreciated, acknowledged, and sought out by others? It feels good. But it is much more important that you recognize, appreciate, acknowledge, and seek out others.

Your theme is "wait to be invited." This has governed your life. While it might be good for choosing pathways, for the most part it is very limiting. Practice inviting, and initiating. Not as a pathway to long-term relationship, but rather, as a practice of intention, of standing in your own light. It is a way of broadening your scope of being invited by Self.

Opening to the One

She is looking for you. She has been looking for you for a long time. She has found you and is waiting for a sign that you're the one. Is she the one? I am the one for you. If you see me in her, she is the one. She can complete you. I complete you, but in this human world I can only complete you physically and emotionally through another. Mentally, time and ego are annihilated through union with the One, provided you surrender. This is the plan for you.

You must open child. Opening is what you are learning now, through saying yes, through trusting, through the Christ Blueprint study. This book opens you wide and gives me the opportunity to communicate with you in ways and with concepts that were previously unavailable to me because you did not know them. Now as you read I can shine the light of attraction on certain concepts or chapters so they become "The One," the chapter you are compelled to study.

You are ready to let go of several layers of restriction. Know and remember that you are held within God's right hand. You need have no fear, yet, we will be pushing all your buttons to excavate all your buried fears. Remember, as circumstances seem to be ominous, trust, say yes, and pull back your power before reacting. This is the same as "don't react" but really is "react consciously" and pull back your power on the fly. This is the same as setting up a straw system where every time you feel fear, the awareness is a straw and you start sipping back your power. We intend to evoke so many limiting beliefs and other unconscious reaction patterns that you will be like one of those stop action movies of a flower, blooming at an accelerated rate.

As you continuously pull back your power from every illusion, you will rapidly become more and more super-powered. Nothing will stop you. Nothing will want to. The only thing that can possibly stop you now is your own illusion of being anything less than God – the Power and Presence of God.

There is no completion. There is only terraforming as your voracious appetite for illusion continues to devour, and thus, become one with Maya.

Here's where the vision of mosquitoes come in. Your power grows as your presence grows and expands. Imagine millions of mosquitoes as the Power and Presence of God acting through you, pulling back your power from every corner of your body and your field, every illusion they can find anywhere. They are injecting love and trust and sipping illusion.

16. Pulling Back Your Power

"Shame is the systematic repression of the expression of God within everyone."

Pulling back your power is a method for disabling the limiting beliefs that have been stored in our bodies for our entire lives. These limiting beliefs are responsible for creating what we notice and how we interpret it. They also shape the synchronistic events around us which bring us choices and opportunities. If we leave these unconscious limiting beliefs in our bodies, we will continue to experience the world that we have been experiencing. By removing these limiting beliefs, we open the door to a whole new world of experience, a world of abundance, a world of love, a world of freedom from misery and discomfort.

If you have tried affirmations and creative visualization techniques, you may have discovered that not much happens. A big reason is that it is difficult to create more fulfilling beliefs without first finding and removing our foundational limiting beliefs.

To pull back your power, you first need to have an experience in which you feel uncomfortable. This might be having a friend not show up to an appointment, or be running late for an appointment yourself and discover that you forgot your phone. It can be receiving a letter from the IRS, or simply noticing that there is a pain in your back. Once you feel the discomfort, you can pull your power back from it, disempowering the limiting belief.

Pulling Back Your Power

Here are 5 steps to pulling back your power, adapted from the book "Busting Loose from the Money Game," by Robert Scheinfeld. Buy this book to get a full book on this method instead of just a few pages.

1. When you feel uncomfortable with a situation, feel for a pain or discomfort somewhere in your body, usually in your chest or abdomen. Know that it is a limiting belief acting to create limitation in your world. Appreciate that you are an amazing being, capable of creating your personal world according, not only to this unconscious pattern, but also through conscious intentions and prayers. Acknowledge that you have created this limiting belief, or at least have chosen to hold it in place so that it can express itself now.

2. Ask the discomfort to intensify. Feel it becoming more uncomfortable until you can hardly stand the pain.

3. Step into your higher nature and command that you pull back your power from this limiting belief. In the book "Busting Loose," Robert Scheinfeld proposes using the phrase, *"I am the power and presence of God and I pull back my power now."* I find that this works very well.

4. As you pull back your power, feel your power flowing into you or through you, and feel the discomfort crumble or dissipate. I like to inhale deeply with the intention of pulling back my power until the discomfort is gone. It feels really wonderful to feel the power returning to me. Enjoy that feeling.

5. Appreciate how amazing you truly are, to have created such a believable illusion in the first place, and then to simply pull back your power from it and have it disappear.

Pulling back your power from limiting beliefs creates a solid foundation for building a new life through intention. If you have tried to use creative visualization, affirmations, prayer, or any other mental creation techniques, but have been thwarted by your limiting beliefs, you may have seen some temporary or unsatisfying results because they were the result of a combination of your manifestation techniques and your limiting beliefs. While you are focused on your affirmations for a few minutes, your body is playing your limiting beliefs 24 hours a day. This is why they are so powerful. Pulling back your power from your limiting beliefs allows your intentions to manifest much more easily.

Foundational Assumptions

There are several conceptual building blocks that may help you to understand this process in order to fully engage in pulling back your power. I have broken it into five basic assumptions. Take them one at a time and be sure you grasp one before moving on to the next.

First, there's a difference between the Earth and the world. The Earth is fixed and definite in form, shape and attributes. It doesn't have any attributes which are any more "important" than others. The world, on the other hand is a world of experience and perception. I live in a world of scientific inquiry, constructing research equipment and searching for Etheric Sludge. This is what is important to me. Your world is different from mine, and as you move your attention from interest to interest, your world can change dramatically. You have a physical world of sensual experiences. You have an emotional world of love and peace, turmoil, anger, and grief. You have a mental world of inquiry, curiosity, and exploration, and of perceived responsibility and stressful overload. Which world you live in determines how you perceive the events that occur around you. You can jump from one worldview to another. For instance, if you find a spider hanging from a thread over your bed, or a mouse on the floor, you might jump into fear, and not notice any of the above experiences. *Your world is created as you assign importance to things or experiences in your environment, including the words and deeds of others around you.* Do you accept this?

Second, the human brain is only capable of registering a very small percentage of everything that goes on around us. To deal with all of this information, the brain needs filters. There are whole collections of things that we don't even notice. Some people will notice one type of thing because it suits their interest or their fear, while other people will notice entirely different sorts of things. We tend to be creatures of limiting beliefs, and because our attention is focused in such limited ways, we perceive only a very small percentage of the possibilities that are otherwise open to us. Earth is a place of extraordinary and abundant possibilities. There are 100 ways you could achieve whatever you want. Some people become billionaires. Others find the love of their life. Yet others find fulfillment in a life of sex and sensuality. Many joyfully spend their lives in curiosity and exploration. Yet others toil and struggle, and find life to be an annoying and painful ordeal every step of the way. So the challenge is, "How do we reprogram our bodies to draw our attention to those things we want more of in our lives?" This is a tough one because it says that *you recreate your world through what you focus your attention on.* Do you accept this?

Third, while on Earth everyone appears to be objectively equal and interacting with other people, in our individual worlds the situation is very different. Your world is all about what you notice, what your body programming draws your attention to. When another person makes a comment, your attention is drawn to those interpretations of his or her words which best fulfill the beliefs and expectations of your world. This means that you are not having a conversation with them, but rather are filling in the details of your world. For example, someone may offer to give you $1 million and you might interpret that as making fun of you. I certainly would. Thus, it is very important to understand that everything that anybody else does or says in your world is exclusively for you, and it has been very tightly filtered and interpreted to fulfill your beliefs and expectations. The saying "don't take anything personally" is very appropriate here. What others intend in their worlds is entirely different from what you receive in your world. What they intend to say is irrelevant to your world. What you receive is irrelevant to their world. You truly live in different worlds. Thus it is a complete waste of time to get hurt or offended by anything anybody says. This is a big one because it says that *your world is yours alone. Nobody else can truly*

understand you, and you cannot understand them. Everybody has
their heavy filters. Do you accept this?

You create your world. This is an especially challenging one.
Everybody loves to blame somebody or something else for their negative
circumstances. If I can't blame my boss, my spouse, or the other driver,
I can always blame my childhood, my genes, or even my Etheric Sludge,
my limiting beliefs. Yet here I claim that you have full and complete
responsibility for your world. They don't! You are the one who is in
control of where your attention is focused. If you let it wander and
repeat lots of old negative belief patterns, that is exactly what you'll
perceive in the world around you. On the other hand, if you practice
mindfulness; being conscious, really conscious of the world around you
and are conscious of your thoughts, your words, and your actions, you
will quickly become aware that you are creating your world. This means
that, *by focusing your attention away from fear, limitation, and disaster*
and toward opportunity and abundance, you will start to notice
opportunities for abundance in your world, and you will be able to start
choosing them, instead of choosing fear and disaster. Do you accept
this?

Fourth. This is perhaps the most challenging concept to understand
or accept. There is a higher aspect of you which is connected to the
intergalactic synchronicity computer, God, or whatever you want to call
it. This synchronicity orchestrates the events which happen in the world
around you. Essentially this higher order intelligence is an opportunity
generator. It is the principle behind the synchronicities that we observe
in our lives. Have you ever discovered that what you were looking for
all along is right in front of you waiting for you to see and choose it?
Open to acceptance of this source of abundance.

On the other hand, have you ever said anything to somebody else
and then thought, "I didn't mean it to sound like that." Then you had that
person react with rage, affection, rejection, or undue appreciation. In this
act you may be observing the intergalactic synchronicity computer in
action. Essentially the other person's higher aspect spoke through you to
them so they could hear what they needed to hear. You said what you
needed to say and they heard what they needed to hear. It was their

higher nature fulfilling the subliminal beliefs within them, and using you to do so. Thus there is a higher aspect of us that is orchestrating our worlds, not only directing our attention to those opportunities that we desire and pray for, but also speaking through others to guide us on our way. Again, it is important to remember, "Don't take anything personally." If you say something that offends or enlightens somebody, but you did not intend to do so, simply assume that it is all about them and their world, and don't take credit for the enlightenment or punish yourself for the anger that they misdirect at you. *Do you accept that synchronicity acts to bring you choices: opportunities that seem to materialize out of the blue?* Another important one: *Do you believe that your prayers and your limiting beliefs guide this synchronicity to help fulfill whatever you are consciously, or unconsciously, asking for?*

The best way to discover this is true is to take two minutes every morning and ask for what you want during the day. Or, if you would like to experience really weird stuff, take 2 minutes to ask to be shown *that this is true*, and then watch throughout the day as amazing, weird, impossible things happen around you. I had an amazing experience with this. I went to my volleyball game and started playing. Nobody on my team could make a single accurate shot. It was an amazingly stupid game because the other team played well and my team botched every point. So I prayed. I asked that everyone on my team play well. The next game my team showed an extraordinary turnaround. We played spectacularly. However, the other team botched every shot. So... I prayed that both teams play very well, and to my amazement, we had a spectacular competitive game. Everybody on both teams was cooperating with my higher serendipitous self to show me what I was looking for. Nobody else had any idea that I had created this extraordinary world for myself. Try it out. Two minutes every morning. Pray for what you want, and watch what happens.

Fifth, you will often discover that, when you pray for what you want, or when faced with the opportunity to choose want you want, often it will conflict with a limiting belief that is embedded within you. Thus you are likely to experience a sudden discomfort in your chest or abdomen or some other part of your body. This is extremely important.

Pay attention to it. ***Praying for what you want, and then discovering where in your body you are cancelling out your prayers allows you to know both the content and the location of limiting beliefs stored within you.*** This is huge. I don't expect you to accept this upon reading it. However, if you will spend 2 minutes praying every morning, then feeling your body for a few minutes afterwards, you will discover that this is true. You will discover that your prayers or affirmations sometimes make you feel uncomfortable inside as what you are asking for conflicts with one or another of your limiting beliefs. Whenever you discover a limiting belief within you, you have a chance to pull you power back from it.

By taking your time with the above concepts, and getting used to each one of them you will then be prepared to engage fully in the process of Pulling Back Your Power.

Since I started this process of pulling back my power from limiting beliefs, I got lots of help from many unexpected directions, not just volleyball games. My 4:00 AM writing practice gave me a great deal of encouragement and guidance along the way. I even got dream interpretations as you will see below:

Pulling Power Back from Shame

4:00 AM

I had a dream that I was visiting a farm and the owner was about to burn down a building to try to hurt a tenant who had her stuff there. I was on my way to warn her when a gang of bad guys surrounded my car. I was in the back seat and tried to drop my wallet under the seat to hide it, but the leader noticed that I dropped something and asked "what are you trying to hide?" At that point I realized that it was illusion and that I could pull my power back from it. I pulled back my power from the discomfort in my solar plexus and the police arrived to save the day.

Noticing the details of how you brought about the problem is important. It can be pure synchronicity, but more often, there are things you do that draw undesirable situations to you. Shame has been one of your big limiting beliefs, a way in which you have self-sabotaged all your life. Continue to observe how you try to sabotage yourself, and how we often rescue you.

What is shame? And where did it come from? Shame is indeed an odd control dynamic. People feel like they have done something which was really bad and brings dishonor to their family or tribe. They get severely reprimanded and sometimes punished by taking their allowance away or excommunicating them from the tribe. They feel an emotion of shame, a feeling of collapsing inside, of not deserving to even be alive. Even dying would not obviate this transgression.

Shame is purely a control dynamic created over the millennia to try to control children, tribe members, and employees. It is extremely effective on most people, regardless of whether what they did was really "wrong" or not. The problem with using shame as a control dynamic is that people internalize it and make up stories about themselves. "I'm not worthy." "I bring dishonor to those around me." "Whatever I do turns to shit." All these limiting beliefs cause people to hold back, to do less or to do nothing in fear of making a mistake. Shaming is a technique to disempower or emasculate people, to make them docile so they can be easily controlled. The result is that they are quiet, obedient underachievers with a lot of internalized, repressed rage. This is what you are feeling over your liver and gallbladder and are using the Infratonic S to alleviate.

The shame is a big "NO" preventing your power from rising above your diaphragm. The result is stagnant energy at your liver/gallbladder. As you pull back your power from this area and realize you are a sovereign being, fully empowered to make your own choices, and determine their actual consequences independent of controlling judgments of others, you become free to express your power in this world. Your liver/gallbladder discomfort will disappear, and your power will express itself through you. You will feel it flow within your body and in your world.

A popular quote by author and speaker, Marianne Williamson, addresses undoing shame at the level of ideas or concepts. She inspires people to see how their shame has caused them to try to be small and relatively ineffective, so they can throw off the limiting belief, and allow the glory of God to shine freely. *Here's the quote:*

"Our deepest fear is not that we are inadequate. Our deepest fear is that we are powerful beyond measure. It is our light, not our darkness, that most frightens us. We ask ourselves, "Who am I to be brilliant, gorgeous, talented, and fabulous?" Actually, the question is, "Who are you not to be?" You are a child of God! Your playing small doesn't serve the world. There is nothing enlightened about shrinking so that other people won't feel insecure around you. We are all meant to shine, as children do. We are born to make manifest the glory of God that is within us. It's not just in some of us, it's in everyone. And, as we let our own light shine, we unconsciously give other people permission to do the same." --"A Return to Love," by Marianne Williamson

Shame is the systematic repression of the expression of God within everyone. 90% of shame is self-imposed. You learned to shame yourself as a child and will continue to apply it throughout your life until you see it clearly and pull back your power from it.

What is "Pulling Back your Power?" Shame is a wonderful example to start with on your own being. Power, raw power, primal power, the power of God to create at will instantaneously, enters the body, enters the physical world through the perineum at the base of the torso, or the first chakra. It flows upward to the 2nd chakra where it is flavored with needs, attachments, rage, etc. Then it flows up to the 3rd chakra where it is often blocked by limiting beliefs that are frequently the product of early childhood shaming.

This power to create is still available to make babies because this function happens at the 2nd chakra. When the diaphragm is cleared of limiting beliefs, the power of creation is free to flow up to the throat where it can be expressed through the voice. This is why we encourage you to speak your prayers, your wants and desires, to command them into existence. Practicing prayer and commanding daily will gradually open the channel from the first chakra up to the throat chakra.

Pulling back your power is standing in your power, feeling and seeing the obstructions within you, and praying and commanding for them to disappear so your power can flow freely. Essentially, you have given control of your power to these early childhood limiting beliefs, which have been put in place by such techniques as shaming. Thus, pulling back your power from limiting beliefs involves seeing that these limiting beliefs are extremely illusory, held in place through residual programming – realizing that they only remain by your choice, and that you are free to choose to have them dissolve so that your power can flow freely from the root chakra, up to your throat chakra.

The "Power" Center

The third chakra, or the solar plexus chakra, is often called the power center, not because it is a source of power, but because it is a source of restriction. It is important to see the difference between real power and worldly power. Worldly power is the ability to insert oneself into a system such that you can say "NO," and thus, can extract favors or money in exchange for saying "YES."

> *"Neither you, Simon nor the 50,000,*
> *nor the Romans nor the Jews,*
> *nor Judas, nor the 12, nor the priests, nor the scribes,*
> *nor doomed to Jerusalem itself,*
> *understands what power is,*
> *understands what glory is,*
> *understands at all."*

--Jesus in "Jesus Christ Superstar"--

Worldly power is the source of limitation in this world because a bunch of bureaucratic controllers insert themselves into systems such that they can say "NO." Then they use this "power" to create personal wealth at the expense of worldwide abundance.

When you pull your power back from the limiting beliefs within you, you discover that you have abundant vitality, abundant health, and the ability to fulfill your wants and needs abundantly without restricting other people's abundance. You can have abundance despite all these controllers who try to restrict abundance for their own personal gain. As long as anyone allows himself to be controlled by shame, there will be controllers who acquire wealth by leveraging that shame. It is not a matter of rescuing these people from the "bad controllers" because the restriction to power is within each person. It is far more effective to teach people to pull back their power from their own limiting beliefs than from controllers because when they recover their own power they will choose to disempower the false gods who say "NO." *Teach a man to fish and he can feed himself for a lifetime.*

Pull Back Power from "I Don't Deserve!"

On the way to the warm, sunny summer. Prepare to receive the rays of the new Sun in full force, burning away the old with relentless, yet steady determination. Feel the joy of liberation, or freedom.

It has long been special to listen to the brook or the ocean or the wind. Now it is special to listen to the flow of blood through your brain, blood through your gut or breath through your chest. As you get quieter all becomes more sacred.

The Infratonic S has done a great job of pulling to the surface hidden ways your body wants to say, 'No. I don't deserve!" Pull back your power from 'I don't deserve.'

You are a King. It is all yours. It is not a question of whether you deserve. That's a judgment. I wish to provide you the entire kingdom. I can't do it without your cooperation. Kingship trumps deserving. There is no limitation, no scarcity. There is only abundance, commanding, knowing that all is given to you. No question of whether you earned it or deserve it.

You have inherited it. It is a gift of your Father. Are you really going to sit in "I don't deserve my inheritance?" Your business is a gift. Your research opportunities are a gift. Your new life in Ecuador is a gift. Your new friends are a gift. They are all the inheritance of a King; your inheritance. Pull back your power from all the fears that come up as you receive the gift of kingship.

With the choice of not receiving, your life can become drudgery. It can shrink or it can grow in joy, abundance, and bliss beyond measure. It is not just you. It is everyone in your world. You are choosing whether your world is to be one of limitation or abundance.

Why would anyone choose limitation? Why have you repeatedly chosen limitation? "I don't deserve!" You have come this far. You have arrived at the top of the mountain, at the Kingdom of Heaven. Choose now to pull back your power from judgment, from limitation, from "I don't deserve." Find these blocks in your body now. Pull back your power, your birthright, and sit at the right hand of God.

Remember to keep breathing back your power. Walk at the right hand of God. Know that life has never been full and complete until now. Walk at the right hand of God. Every meeting, every partnership, every friendship, every project is you meeting God and receiving and fully accepting your birthright.

Step up and be the King. Open your arms and receive the Kingdom which is your inheritance. You were called. Answer the call fully, completely, without reservation.

Incarnate fully now and feel your feet, ankles, and legs. Feel your presence fully occupy your body. For your new body has inherited the New Earth. This is your gift and your birthright. There is no place for "I don't deserve!" Pull back your power from it fully and completely.

"I deserve all the gifts my Father gives me because He loves me so totally and completely." Feel the love of your Father now; your physical father and your heavenly Father. Allow your physical father to be the full embodiment of God the Father, and receive him fully. Accept that he loves you fully and completely. You are held, supported, and loved.

Every gift is a gift of love. Every request you make is an opportunity for your Father to show you how fully and completely He loves you.

Ask anything in his name and it shall be granted. Your life and every life in your world shall be blessed and uplifted every time you ask with an open heart. Be open completely to receiving every bit of abundance that your Father can offer.

Prepare to be inundated, blessed beyond your wildest dreams. And know that, as you receive, your life and all the lives around you are enriched beyond measure.

Pull back your power from "I don't deserve." Do it now. Receive the Kingdom of Heaven now. Open your heart, mind, body, Spirit, and your world to Love – to all the blessings your Father has in store for you. Turn on the shower and allow the warm water to flow. As you open the faucet more and more, feel the joy and abundance mount. Feel the abundance fill your entire world and everyone in it.

I deserve abundance. I deserve the love of my Father. I deserve the love of every researcher, friend, customer, supplier, every plant, animal, crystal, every gust of wind, every wave, every clap of thunder, every drop of rain, every ray of sun. I am loved by my Father. I deserve it all!

Pull back your power from "I don't deserve." And know that you are pulling back your power for everyone in your world. The limits shall crumble and Abundance – Heaven – shall descend upon you and the beloved planet Earth.

Embrace your kingship, your inheritance, your stewardship. And pull back your power from every corner of your kingdom, from your world. Do it now!

Pulling Back Power from All Earthly Limitation

What a blessed morning it is. Relax and know that all is well. Nothing should concern you ever. Know that every minute is full of opportunity; every interaction, every chunk of silence. Be open to the doorways of opportunity. Remember to choose. Follow the feeling. Pull your power back when you feel discomfort or fear. Feel yourself in anticipation. Feel the fear in front of the knowing. Fear of success beyond your wildest imagination. *Why the fear?* Because of the opportunity. *What is the opportunity?* Your fears keep manifesting the illusion of failure. Relax and become the success which lies just beyond the fear.

I am the power and presence of God pulling back my power from the fear that I'm not good enough, that I don't deserve. Who am I not to be brilliant, successful beyond measure? I pull back my power from the fear of punishment for brilliance and success. I am safe, 100% safe. The past is melting away. I pull back my power from all the punishment for brilliance and service.

The Infratonic S has opened the door to pulling back your power from all repression of all heartfelt service in the world. Pull back you power now. That's right! Feel it in your entire body. Feel the power of the infinite returning. Know that you are changed in the blink of an eye, as is everything and everyone in your world.

This is a huge door you are walking through now. Keep sipping back your power. Feel the indomitable joy emerging. Pull back your power from *ALL EARTHLY LIMITATION*. Keep in mind and heart always, "I am joy and light." Continue pulling back your power before you go to bed and when you wake.

Completing the Trauma Complex

Kelly is, as you have thought, traumatized. Her vagus nerve is blocked with Etheric Sludge. It is melting with time. So much self-blame and shame. So many buried secrets. It feels like they are secrets from this lifetime. Yet they are from past lifetimes. Things she had died with. Things she had taken to her grave. Guilt, remorse, shame.

She can't speak of feelings because she can't put words to the past lives. She remembers all about the main traumas of this lifetime, and all the details that led up to her father's departure. The abrupt departure and every little interaction she had with each parent that led up to it. She has replayed every event to find out what she did wrong. She feels that she knows that she is responsible for the disaster. Yet she can't find the cause.

It is a case of several lifetimes coming to a singularity, to a similarity of circumstances. It is so important that she be able to experience this pain of so many lifetimes because it is her opportunity to release them all at once. She has processed so much of it, and released some, but the bulk of it must be processed together. It is one huge lesson. One giant opportunity!

She has made the choice to release it all in this lifetime. The complexity goes further. She did not have all of the needed traumatic ingredients from her own life. She is an empath. She has borrowed trauma from her parents' lives, from siblings, from friends, from school enemies. She has been a sponge for other people's traumas, whatever will help complete the picture of the trauma complex that she has repeated lifetime after lifetime.

Her trauma complex is now complete. She needs to draw no more pieces from the outside. She is ready for the convolution, the emerging of a higher elegant reorganization of her psyche – from chaos to a diamond of extraordinarily high radiance.

She is afraid. She has spent so many lifetimes trying to collect all the pieces and bring them together. Now that she is on the verge of emergence, she fears: What will it be like without this pattern of remorse and shame? Can she live with the loss?

Her purpose for 100 lifetimes has been this moment coming soon. The exquisite reconfiguration – literally the Transfiguration focused on in the Bible. She is ready to undergo this for herself and for all of humanity.

Yet…On the other side, she faces the unknown, the emptiness, the void, the uncertainty. When she lets go and allows the Transfiguration, her entire purpose, lifetimes of purpose, will be gone. She knows the moment is at hand. Like Dorothy and her magic slippers, Kelly can tap 3 times and the Transfiguration happens.

One step through the chaos and through the pain, but what will be left? Will her life be pointless? Will she wander through life from then on as an empty shell, cut adrift, purposeless, lost? This huge fear is the only thing keeping her from tapping her shoes together and initiating the ride of a lifetime, of 100 lifetimes.

When she lands, she shall be held, loved, and exalted by Spirit. Nobody in this world will know that she has spent the hundred lifetimes in sheer agony for this huge purpose. Nobody will care. She won't even care when she lands. It will be like she lands in a huge pile of sawdust. All of the pain and trauma of lifetimes left in a big pile, dead and lifeless. There will seem to be no point to the hundred lifetimes of suffering. And yet, in that moment of ecstatic Transfiguration and reunion, she will see how it all fits together. She will know why the sacrifice was so important for her and all of mankind.

Then she will emerge from the ecstasy. Such extreme love, joy, and bliss cannot be sustained in one little body in one lifetime at one point in time. She will stand up, dust herself off, realize the huge liberating freedom of the life ahead of her, and take her first step forward knowing that it will be perfect. She will know that every step she has taken over the last hundred lifetimes has been perfectly orchestrated to lead to the moment of Transfiguration. Then it is done and there is nothing to do but to take the next pointless, but joyful, loving and blissful step.

A life of empty, pointless steps of joy, love, and bliss. She fears the "empty" and the "pointless." Her friends and family will not appreciate her achievement. After 3 steps, she won't even appreciate it. She will have only this pad of paper to remind her that it ever happened, and by the 3rd step, she won't even care about that.

Trust

You have the opportunity to grow in big ways. As you discover emotional discomforts in your body, pull your power back.

What is the pain in my chest? Is it betrayal, hurt? That I trusted and now feel deceived, betrayed?

Your life has been one of trust. You need to trust in order that you feel whole and loved in this world. If you don't trust, you can't experience the feeling of being fully alive. You can't feel a part of this world. You need to give the world a chance to love you back. If you don't give the world an opportunity to apply free will, you can't be either disappointed or validated. Trusting is the way to give the world a chance.

Know that the rest of the world is you, acting as mirrors for you. This is a golden opportunity to pull your power back from illusions; the pain and suffering you have chosen now, over the years, and over the millennia. Know that what you feel now is what you would have felt time and time again. See this clearly. You have created your suffering. Feel the pain in your chest, the burning. You have created this feeling as punishment for misdeeds long forgotten, and irrelevant to the present.

I am the power and presence of God in this world, meeting other versions of myself, holding myself accountable for all transgressions against me. In reality all is forgiven. However, I must forgive myself for everything I've ever done and will ever do to me.

I am the power and presence of God. I'm examining the illusion I've created of suffering. I am suffering and blame myself for past transgressions against others and in punishing myself by having my world punish me for what I, in my created illusion, have come to believe I deserve. It is illusion, the belief that I deserve to be punished, which is the illusion which I am pulling my power back from now. I feel deeply how I have punished myself over and over again. I feel the burning…forgive me! Forgive me for doing these deeds. Forgive me for punishing myself over and over for elusive wrongs I have blamed myself for many times in the past. I have put myself into the position in which I feel taken advantage of. I have done this to get myself in touch with my own judgments against me. I am the power and presence of God. I am also the love, joy, bliss, abundance, and vitality of God, choosing to experience limitation, pain and suffering, so I can see that I am creating my own suffering. I can choose to pull my power back from these illusions now.

I am the power and presence of God. Having pulled my power back from all illusions of suffering, I am now experiencing myself as whole and complete. I am forgiven! I am the world around me and I love me fully and completely.

"I am forgiven. I am joy. I am love."

17. The Alchemy of Creation

"Plant seeds of love, joy, and bliss and
grow a productive crop worthy of harvest."

In their performances, the Qigong Masters demonstrated such things as breaking bricks, making body parts super hard or super strong or bending steel. This is most impressive, particularly when seen in person and up close. Qigong exercises associated with this generally focus on the lower Dan Tian, which is located in the lower abdomen about an inch below the navel. During meditation the intention is focused there so that Qi will build up there which can be utilized to perform these marvelous feats of power. This provides the impression that Qigong is about cultivating power.

From the movies, and from many of the Qigong practitioners I've come across, there is definitely a huge emphasis on demonstrations of power, often associated with intimidation. Certainly there is something to be said for the saying, "When you have a hammer, the whole world looks like a nail." Certainly, for those who view the world as entirely physical, as modern medicine views the physical body, an emphasis on these physical demonstrations and on power and intimidation is inevitable.

The Qigong training I went through involved some of this physical Qigong called Jing gong, but it also involved a great deal of emphasis on Qigong, the cultivation of Qi and heart essence. This is very important for the healing arts because while cultivation of Jing cultivates strength, hardness, and firmness, cultivation of Qi cultivates fluidity, action, movement, and flexibility. Connecting with trees, with sky, with water, and with Nature is very important. In the writings I've included in this book, this is presented as moving away from separation and into Oneness, from limitation into Abundance. From the state of Oneness and Abundance, most anything becomes possible, particularly remarkable healing at all levels. Healing is becoming one with the whole, which encompasses letting go of separation and limitation.

While in China, I didn't get much instruction into the 3rd kind of Qigong, which is Shen gong, the cultivation of Shen or spirit. On the other hand, perhaps the training was all there, and I wasn't ready to learn that level. The teachers emphasized that quiet "sitting Qigong" is very important in the development. Since language was a barrier for me and for the 50 or so doctors and therapists who accompanied me to China, teaching physical and movement exercises was much easier than teaching quiet internal cultivation exercises.

In one demonstration several students had me lay down on the bed and they held their hands very close to my head for 10 or 15 minutes. I received very little explanation, and when they were done I felt about the same. However, that night when I went to bed and turned off the lights, the room was very dark. As I closed my eyes, the room was filled with a bright light. I was stunned. I open my eyes again and saw that the room was still dark. Nobody had entered the room and turned the lights back on. I closed my eyes again and the room was again filled with light. Amazingly, they had filled my head with light. I still don't know what their intention was. Perhaps it was to help the foreigner see the light. Or maybe it was to plant a seed in me, either a conceptual seed, a possibility that I would continue to consider over many years, or a physical seed, the cultivation of an awakening of a part of me that I had known nothing about. I suspect that these students gave me a huge gift that stretched beyond my comprehension.

Below, I describe the 3 seeds of human consciousness which the Chinese call the lower Dan Tian, the middle Dan Tian, and the upper Dan Tian. Each of these lies asleep and dormant within us until it is awakened, either by our own introspection and meditation, or through the intervention of other people, or Spiritual Beings.

As the upper Dan Tian develops and flowers, our analysis and worry, our thinking-but-not-knowing, transmutes into an intuitive knowing in which analysis and worry are no longer required or desired. They are recognized as Etheric Sludge and discarded.

As the middle Dan Tian develops and flowers, our limitation and isolation are seen as unnecessary sludge, and transmute into oneness and abundance. Our knowledge transmutes into loving wisdom.

As the lower Dan Tian develops and flowers, fear and separation are again tossed into the Etheric Sludge bucket, and transmute into blissful connectedness. The need for power to impress, intimidate, and acquire is replaced with the power to manifest abundantly through synchronicity and Grace. Need vanishes from our world.

Claiming Wisdom, Love, and Power

4:00 AM

There are three locations in the body that correlate to the 3 stages of the creation process. The **head** is the center of creative visualization, the first stage in the alchemical process of creation which begins by visualizing what is wanted. This can be governed by thinking within the accepted knowledge base of society, or by Wisdom, knowing within ourselves what feels right. Learning to trust ourselves is the pathway to Wisdom.

The **heart** is the center of vitalizing, the process of wrapping the intention in vitalizing desire. This is an important stage because the magnetic stuff of desire attracts the physical to the visualization to begin the process of materialization. In most people it is governed by fear, drawing to us that which we want most to avoid. Love vitalizes, fear causes contraction, stiffness. Pulling our power back from fear is the pathway to Love.

The third stage of alchemical creation is described as sending the thought forms out into the world and releasing them. This is desirable from the standpoint of releasing ongoing wanting related to the creation of the thought form. When we continue to want something strongly we often build a big ball of desire that plays havoc with our etheric template causing problems with our physical and emotional health.

This third stage of creation involves the creative center located in the lower abdomen, sometimes called the **hara,** lower Dan Tian, or the womb space. This is a very important and little understood organ of the energy body. In creation of new human life, the seed is planted in the womb space, and a new embryo forms and grows into a human child. While this is a creative use of the womb space employed by most of humanity, this creative space can be used for much more. Claiming this space is a pathway from limitation to Power, the Power to create our lives exactly as we wish.

Whatever we successfully put into the creative space is created in the physical world. However, most of our creation in this regard is unconscious. We go about building with our words and with our hands, unaware that our hidden, unconscious limiting beliefs are often creating obstacles to what we believe we want to create. Many of these limiting beliefs live in the area of the diaphragm and solar plexus and strongly control what goes into the creative womb space. Thus lots of flavored creations go into this space, which often contain thoughts like "I can't do that." "I shouldn't be that." "It is wrong to create that." We wonder why we keep procrastinating or can't hold to a diet or are constantly late to appointments. In the creative process, we keep flavoring our intentions with hidden limiting beliefs.

We are taught in industrial society to stay in our heads and live in our heads. There is some movement to encourage us to move into our hearts. However, there is very little teaching on moving into, or taking charge of the creative space in the belly. Society discourages us from doing so. Moving from unconscious to conscious control of this space is moving from restriction and limitation to Power.

What happens if we are not consciously aware of, and don't plant and cultivate, seeds of our choosing in the belly? What's there? We could say weeds. That's what grows when we don't cultivate a garden. Interestingly, a garden starts out as all weeds before we decide that we want to turn it into a garden. The first thing we need to do is pull out all the weeds. The second thing we need to do is be sure the soil is full of lots of nutrients that will nurture whatever we plant. Third, we plant our seeds. And fourth, we don't just sit back and wait. We continue to water, nurture, and pull weeds.

Reclaiming the Womb Space

We know where the creative space is. In both men and women it is in the lower abdomen just below the navel. The next challenge is to get in touch with this creative space. We must see it or feel it or otherwise somehow know where the weeds are and what this creative space needs for nourishment. The most obvious physical symptoms of weeds in women are symptoms of pelvic inflammation, and in men, prostate inflammation.

The easiest way to get a handle on what is currently in the womb space is to look at the most powerful plants and weeds in the womb space. These relate to our hidden limiting beliefs regarding sex. "I am unattractive" might be a popular weed, whether I am too fat, too skinny, too short, too tall, or simply, despite all appearances, a core belief that "I am unattractive." With a seed like that planted within us, we will find that we create a world that finds us unattractive. Similarly we can have weeds that create "I draw people who are violent, disrespectful, or cruel." or "I am rejected over and over again."

What happens if we leave our garden unattended? Others plant *their* seeds in our garden; television advertisements for the latest junk food or gadget or supplement to make our lives better. If we are unconscious of our creative belly and others plant these seeds we will find ourselves obtaining the products and services they offer, even if we don't want them or they don't serve us. Network marketing is so successful because salespeople personally plant seeds of desire in the creative space of the customer.

Through our genetic heritage our creative space is planted in weeds from patriarchal and matriarchal traditions. Men carry weeds that implant violence, hatred, resentment, and attitudes of servitude in the womb space of women. Over the centuries this has become a major source of pelvic inflammation. From an ancient era of matriarchal dominance, women tend to carry weeds that can cause them to implant seeds of control and emasculation in the creative space of men. This contributes to men's tendency to fear commitment. These ancestral tendencies are a big part of why harmony between the sexes is often such a challenge.

Women can criticize men for being abusive and violent, and men can criticize women for being controlling and emasculating, but the problem in both cases is not outside of the individual. It is within the creative space in their bellies. Cultivating this creative space, pulling any weeds which interfere with a world of nurturing and uplifting people around them can enhance relationships.

Why are we so unconscious of this creative space in our bellies? Why do we find it so difficult to even find our own garden? Because we have been programmed not to look for it.

Over the millennia it has been to the advantage of others to have us unconscious of the Power that controls us. This allows them to implant their own control dynamics within us to get us to do what they want us to do. A reason men might implant violence and belittlement in women is so women will find it too painful to go into their bellies and take back their own power. Similarly women might fill men's creative space with emasculation so men won't feel worthy of taking back their own power. When a garden is unpleasant to visit we don't go there, and others can have their way with us.

It is a sad state of affairs when people belittle others for the sake of control. Yet this happens in so many aspects of life. We are filled with contradictions that undermine our wisdom, fear that undermines our love, and belittlement that undermines our Power. When we start appreciating, supporting, and nurturing those around us, everything shifts. We must make this change at the level of our creative center. It has nothing to do with changing other people in our lives. The change must happen within.

Growing Quickly Wise

R.:   I was at Starbucks recently drinking my tea and doing some writing when a woman sat down nearby to read a book entitled "Growing Slowly Wise." She explained that she believed she was not wise and wanted to be, and that it was really hard to be wise. I countered, "The only thing that is preventing us from being wise is all the clutter of other people's ideas within us." ["The belief taught to us by our society, and particularly our school systems, is that the only valid knowledge is that presented by others, and particularly, experts. If we can pull our power back from the idea that only those outside of ourselves are legitimate sources of wisdom, we can stop the automatic tendency to repeat what we have learned from others. Then and only then are we able to quietly go inside and feel what is true for us. This is wisdom. Instant wisdom. Growing SLOWLY wise is a waste of time."]

During this conversation I realized I was feeling immense sadness in my heart and saw that she was crying. It looked like she had been crying for a long time. I didn't ask for her story and she didn't share it. After the conversation it took me more than an hour, and the help of the Scalene Light, to be free of the sadness. When it was gone, I discovered a cord between her 3rd chakra and my 3rd chakra. This was most intriguing. My attention had been drawn to the heart pain I was feeling, and I had been unaware of how connected we had become through the 3rd chakra. As I pulled out the cord from my 3rd chakra I could feel that it was connected to my 2nd chakra creative space.

This was most intriguing because I had just attended a workshop on Womb Wisdom, and found myself suffering from an inability to "get into" this creative space. I continued to pull on the cord created during my conversation with this woman to see what I would find. There were all sorts of devices, gadgets, weeds, and cloaks surrounding my creative center, clutter left from lifetimes of neglect. I instantly saw how attractive I would be to a woman who wanted to manipulate me. The control mechanisms were already all in place within me. All she would need to do is link up to them and start pulling the strings. With this, I had discovered the residual baggage in me from the ancient matriarchal society through which women easily controlled men. Women who unconsciously control by belittlement would probably attract men with precisely such tendencies as mine. I continued to pull out everything I could find from my creative space that looked like a control dynamic.

I saw that I had been creating and recreating the world around me and my experiences by leaving control dynamics and other weeds in place in my creative space. I realized that I can plant seeds of independence, of confidence, of repelling those people who would want to control me, and attracting those people whom I will find fulfilling and growth oriented.

The Womb Wisdom workshop had planted the seed in me that I could become aware of my creative space and learn to cleanse and nurture it. And this woman who was in pain and looking for support allowed me to connect with my creative space and enter the road of conscious weeding, tending, and nurturing.

There's obviously more than just relationship stuff in this creative space. There is massive stuff from the media, from advertising, from all sorts of sources that we need to weed out in order to clear the creative space so that it belongs to us rather than to those whom we have allowed to control us.

Thus we find ourselves back at "Slowly Growing Wise." Our patriarchal society teaches us that we can only believe those things that experts tell us; that anything that we spontaneously feel to be true should be rejected in favor of the opinions of those with M.D.'s and Ph.D.'s. One of the first weeds we need to pull out of our garden is the belief that we can trust what other people say as true. Believing this opens the door for others to plant *their* garden in our creative space. We need to get in touch with our own knowing, and gauge everything we hear and see based on how it feels within us. We are taught that we are ignorant and need to rely on others for truth. In truth, we are innately wise.

If we simply find that weed and pull it, we find that we can start looking inside for the truth. We can start feeling inside for what feels right and what feels wrong. We can then start pulling all those weeds that don't belong in our garden.

Anything that feels like fear is a weed and can be pulled. Anybody, any article or newscast that speaks of things that evoke fear in us can be pulled from our lives because it is planting seeds of fear and illusion, and opens the door for us to be controlled. Anything that contributes to anger, resentment, or clutter can be pulled as well.

We have been planting new seeds throughout our lives. Once the garden is weeded and we become determined to keep it weeded, we are finally in a position to plant seeds consciously and to grow a productive crop worthy of harvest. We can plant seeds of love, joy, and bliss, seeds that will lead us to appreciate others and the world around us. Appreciation causes the value of the world around us to "appreciate," to become more valuable.

As we plant more seeds and become aware of their growth, we become more aware of the weeds that remain, and more eager to pull them out relentlessly.

18. Diet

An Ideal Diet

I was searching on the web for interesting activities and found a website named Meetup.com. Here, I discovered that I could type in my interests and it would tell me about specific events where people in my local area were getting together to pursue that interest. For instance, I typed in volleyball and discovered that there are games going on every night of the week in local parks near my house. This has been wonderful exercise for me.

I was poking around the Meetup.com site and found a group called "Raw Foods Orange County" and signed up because I had been interested in learning more about a raw food diet. After I attended a couple of raw food potlucks which were fun, I received an announcement that said "Raw Food Meetup in Ecuador." Something in me just came alive, and I instinctively clicked the "YES, I'm Coming." button.

Four of us headed to Ecuador for a two week stay and education on raw food. It was my first trip to South America. I learned a lot about all the different vegetables you could chop up and put in a bowl, and what interesting and health promoting spices could be added. The dishes were delicious. Plus, I learned to add antioxidants to the food, things like cinnamon, cayenne pepper, ginger, and goji berries. One additional, most interesting addition was probiotics. This is a white powder that comes in a capsule and contains all or most of the digestive bacteria that cultures a person's intestines and digests food for us. This good bacterium continues to live in the intestines. When people take antibiotics, these powerful drugs kill off most of this good bacterium, making digestion much more difficult. Until the digestive bacteria are restored, digestive problems result and there is poor assimilation. After about a week and a half of eating what the other 3 members of our group were eating, I noticed that I was having very little elimination, so I asked during my morning writing about what I needed to include in this raw food diet. The following is the answer I got:

4:00 AM

The qualities of an ideal diet for you are many. Those illustrated here are:

> *Eating less.
> *Eating biodiversity.
> *Eating where it is grown.
> *Being among wonderful people.
> *Eating enzyme-rich, living foods.

Deficiencies of this diet are low fiber and excess fat in the form of coconut meat and almonds (even though the dishes are delicious and nutritious). Almond yogurt contains far more fat than animal yogurt. Goat milk yogurt is very healthy, fresh, enzyme rich, and nutritious. Almond yogurt is a great source of certain enzymes, goat's milk or goat Keefer is a great source of different enzymes. The probiotics you are sprinkling on the different foods become a source of many different families of enzymes. *[Probiotics are available in the refrigerated section of local health food stores, and contain as many as 50 or more different kinds of digestive bacteria. You can sprinkle them on any foods so that if you are deficient in any bacterial flora in your gut, they can be seeded and begin to flourish and become an important part of your digestive system, increasing your ability to tolerate foods and decreasing gas and other byproducts of poor digestion.]*

Why do I have so little defecation? An intriguing mystery indeed. Certainly, your reduced fiber intake is a major cause. Also digestive probiotics do a better job at assimilating all foods. There is little left. Many carbohydrates are broken down by the enzymes into H_2O and CO_2. Some of these enzymes transfer this energy directly to the body. *[To learn how these enzymes might transfer this energy directly to the body, be sure to read Chapter 25 on Kirlian photography and the mystery of what causes air to become luminous.]* Many fibers not normally considered carbohydrates become digestible with these enzymes. This

258

means that fiber has disappeared. Whole grain oatmeal with husk will help. Chia seed, sprouts, mung beans, black lentils, kale, spinach, carrots, beets, and krill oil will also help.

These super enzymes you have been sprinkling on the food can easily help feed a hungry world because all of life is enhanced. The body is vitalized and energized. Parts of foods normally wasted become nutritious. Because less does more, less food is sufficient to feed the body.

Some of these enzymes break down proteins to provide amino acids. These enzymes are hugely important for the starving people in Africa and North America. These enzymes allow people to choose when they leave their bodies. This is the answer to the US health crisis. The key is cultivating them in a variety of food stocks as wide as your diet. *[Because this digestive enzyme powder has so many digestive bacteria in it you can sprinkle it on virtually any food that you would eat and those particular bacteria needed to break down that particular food are cultivated in the intestines as that food is digested.]*

For instance, a blend of nightshades fermented with probiotics provides the body with the needed probiotics to digest nightshades. There are probiotics that eat BT toxins and probiotics that eat radioactive waste. Their predecessors already exist. With prayer they will spontaneously spring into existence, or rather Gaia and the Heavenly Hosts will call them forward.

R.: If you are heading to Ecuador to experience this marvelous diet, you should know that the Ecuadorians eat mostly beans, rice, corn, and chicken with a little beef and pork and occasional cooked vegetables. You don't get the above diet simply by traveling to Ecuador. However, the town of Vilcabamba did have a raw food potluck while I was there. A small group of foreigners have imported this raw food tradition.

19. Speaking with the Land of San Pedro, Vilcabamba

When I first visited Vilcabamba in January 2013, I made preliminary arrangements to purchase a piece of land which seemed very special to me. It was at the very top of the Vilcabamba Valley and the headwaters of the Vilcabamba River. I didn't know why I was purchasing it. It just felt right. Three months later, all was ready for the purchase and I was heading back to Vilcabamba to sign the papers. Along with the land, I was inheriting a tenant named Catherine who lived on the land. I sent her an email that I would be arriving in a few days to arrange a time when we could meet. She wrote back:

> "Hi Richard,
>
> "I'm so happy for you that you bought this land. The energy here already shifted again. But at the same time many questions appeared to me and I would like to get to know the answers. I'm wondering. Is my "caretaker" mission here accomplished? What is your next step? Do you have any plans to create something on this land? (I would love to help)
> "I'm doing a "San Pedro" ceremony this coming Friday to ask the Spirit of the land for directions regarding the land and "the ceremonial place." I'm looking forward to hearing from you and I'm wondering if you can join us."

I knew next to nothing about the San Pedro cactus and what a ceremony might entail. My experience with psychedelics was virtually nil, and to discover that I was being invited to what I thought was a psychedelic party on my own land was quite a surprise. Nonetheless, it felt right. My intention was to attend the San Pedro ceremony, but not to take any San Pedro, just to observe. This is what scientists do, and how I felt safe.

Nonetheless, when I got to the ceremony it seemed really safe and harmless. The shaman, a young Westerner, was quite open and explained everything. He said prayers for all the people in the room and for the San Pedro. All seemed very comfortable, so I asked for a beginner's dose, and he gave me about an ounce of the green liquid, about a quarter as much as other people were getting. It tasted sweet and earthy.

I didn't feel anything for a couple of hours. Then I started feeling strange muscle twitching and contraction in my legs. I laid down and enjoyed the sensation. Then Catherine started drumming and singing what sounded like a Native American prayer song. I found myself dancing and singing along. About a half hour later, I found myself scribbling wildly on my pad of paper. The land seemed to be telling me what it wanted. And as it turns out I was the only one the land spoke to that night. Here's what I wrote:

4:00 AM

It has been a new birth of freedom for Catherine, San Pedro, Vilcabamba Valley, and the Earth. The old world has died. Old traditions must die with it. New traditions, never before seen or heard within the language of the past, are emerging. Joy of a new birth, of a paradigm of peace and abundance, spreading from here to the world. There is no end to the beginning of the New Earth.

It is a single focus. Choose one: ayahuasca, tobacco, incense, marijuana, San Pedro cactus. For a new tradition to be born, the old traditions must all die. You must choose one. Catherine must choose one. This land must be about the new birth of a single voice. Humanity is a part of nature. It is not about the human nervous system, it is more about human experience. It is all about a new birth of Community, here and now, in this very moment.

There is only one leader, one focus, and one new teaching spreading around the world. Let it be born tonight. Choose only one voice and speak loud and clear. It is feminine voice to a masculine embodiment of

God. Embodiment is feminine; Yin within Yang, Yang within Yin. It is a world without ending and it begins here and now.

The new energy is poured into the land. Flows down from here to the old ceremonial grounds. The old is being withdrawn. "I withdraw the old. I call back my power from the old." Continue to pull back the past. This is the new initiation.

The seed, the birth, the new! Those who rely on lots and lots of San Pedro to feel are the past. They must be left behind, excluded from the rituals. Richard and Catherine speak with a single voice. Gentle assistance from the San Pedro cactus, (only small doses of San Pedro), catering to those who want to ground and open. It is NOT for those who want a psychedelic experience. Leave them behind. There is always safety in small doses. Focus on knowing community with Gaia. There will be criticism from shamans and those seeking bigger and better psychedelic experiences. This is not our pathway. This does not lead to healing, and this does not lead to community.

Facilitating the release and healing; facilitating dredging old embedded stuff, bringing it to the surface, and providing pathways for easy conscious release and growth. It is the Feminine and Homeopathic application of San Pedro that generates much result.

Comments for Catherine, after breakfast: The way of psychedelic tourists and residents is a dead-end for us. They crave individual growth, individual intense experiences, individual visions that they can tell others about. Any of this will dilute our mission. Catering to the healing you and your daughter were doing is a strong pathway. Catering to establishing community with Gaia is a strong pathway.

For this to work, you need a strong voice and strong determination. You need to understand that you will seem to get criticism from all sides as you step on people's toes for not upholding their illusions of the past. Continue walking forward. More and more will follow. Psychedelic tourists and residents will fall away. Do not allow anything you don't like. This is not about accommodating egos. This is about living your truth. No tobacco, no Palo Santo. Nothing other than what brings people together around your path to healing yourself and the planet.

San Pedro seems perfect. Its focus is on community with Earth and with others through alignment with Earth. Ayahuasca is focused on the individual's relationship with an abstract God. San Pedro's pathway is community with nature. If you offer more than one pathway, you are catering to psychedelic junkies. This is not your pathway and is not the pathway of the land. You can do ayahuasca elsewhere.

The old house must be restored. It is the home of Señor Carpio, famous centenarian. This will draw people. However, it must be presented as a solid historical Museum devoted to healthy eating, to the past and to the future. Wisdom lost is regained!

[I was awake, calm, and relaxed for the rest of that night and throughout the next day. No sleeping for me. But the next night I slept soundly, and for 10 or 11 hours. Over the next several weeks I found myself writing frequently about the land and what it wanted:]

Rebuilding the Temple of Sacrifice of the Land in Ecuador

Over time, a cycle, a resonance, will be established in which the energies pulsate on a weekly basis. New moon, one half moon, full moon, one half moon. Not exactly every week. Clearing after each ceremony will build this pulsation such that energy and clarity are high on ceremony day. If no event on ceremony day, a brief clearing with the Scalene Light is sufficient to build resonance.

When the Ether is called, an inward flow begins toward the temple and expands spherically. Then at culmination of the ceremony, clearing dissolves the sludge, creating a void, a vacuum, adding strength to the draw, amplifying the draw. Thus the peak of clarity will occur after the ceremony. The ceremony is the initial initiation of spirit. The fact that visions may continue for some hours doesn't change the ceremony.

Drinking the San Pedro brew, letting it evoke for an hour, then clearing, will provide the clearest, deepest, most accurate visions, and help avoid bad trips where people sit in their own Etheric Sludge and that of others.

Rebuilding of the Temple of Sacrifice. A permanent roof is not good at this time as the Sun must shine upon the ground after each ceremony. Sacrifice must be cleansed from this land. Without regular cleansing, sacrifice will want to re-materialize at the ceremonial site. Any shaman who evokes fear or limitation of any sort is to be denied access to the ceremonial site. The energies will repel, but eons old patterns are hard to break. Celebration of abundance and vitality, not limitation and sacrifice, is called forth. This is a key component of the sludge to be cleansed.

The ground must be highly conductive of the etheric emanations from the mountain. The hill acts as a lens focusing the emanation. *What conducts/inhibits these emanations?* It is more than that. What conducts and focuses the desired emanations, and disperses and attenuates that which is to be cleansed? We don't want to seal in anything. All should be open to flow. The soil is best. Rocks of the right kind can be good, and can be shaped like a single lens. Cover the foundation in soil and pull weeds regularly. Part of the ceremonial preparation is purifying the lens. A cement cap would be a problem, blocking the cleansing process. Cement beams and posts are okay. Excess steel should be removed.

Once the land is purified, by December 21, 2013, a new dome structure can be built. Existing building materials will be removed. New foundation will be laid. New roof will be built – dome of particular materials will focus the emanations. Wait for it to emerge.

20. The Etheric Template and Embedded Etheric Sludge

A Disrupted Etheric Template

When I first began using the Infratonic in 1978, I did so in clinical work with Dr. Fu. One of our patients was a quadriplegic. She had suffered a neck injury, spent 5 months in the hospital. When she came to us she was paralyzed with no feeling from needle prick below the neck. Dr. Fu explained to her that the effectiveness of acupuncture in treating paralysis decreases with time, and after 5 months of hospital recovery without significant progress, the likelihood of recovery was low, perhaps 30 to 40%.

As Dr. Fu's assistant at the time, it was my role to remove the acupuncture needles and apply acupressure according to her instructions. In this case, I was to apply pressure to Shelley's lower neck and back below the area of injury. Her muscles were extremely tense everywhere and she was sensitive around the injury site. I thought that the muscle spasm might be compressing the vertebrae at the injury site, blocking the flow of blood and slowing recovery.

After the first month of treatment she felt sensation to needle prick (Dr. Fu's method of charting progress) in her shoulders and a few inches down her arm. It was painfully slow progress. It was hoped that, if she could gain the use of her shoulder, she might be able to operate a motorized wheelchair. I thought the Infratonic might help her muscle stiffness. This was the first time we had sent a patient home with her own Infratonic. It was a month before Christmas. We suggested that she use it on her neck in the hope that it would not only relax the neck muscles, but possibly increase circulation at the injury site. She applied the unit to her neck twice a day. Her progress was rapid. As each day progressed I could feel her muscles softening during my few minutes of acupressure. She was happier because the Infratonic relieved much of her nagging pain and uncomfortable tension.

After a week of regular Infratonic treatments, Dr. Fu made another needle prick test and discovered that the sensation to needle prick had moved down past her left elbow and down almost to her right elbow. She had also gained sensation another inch down her chest and 2 inches down her back. On her last visit before Christmas Dr. Fu tested her again. Not only did she have sensation down to her wrists, but a nerve reflex had developed on the bottom of each foot and there was sensation to needle prick in the right foot. This was a very positive sign, indicating that the nerves to the feet were intact, and that full recovery was possible.

Shelley had been using the Infratonic at the site of injury several times a day throughout this period, and while her muscles had relaxed, we did not attribute any of the sensory improvement to the Infratonic. Unfortunately, Shelley found the shape of the Infratonic transducer awkward when she applied it to her neck, so I built another transducer, one that had all the electromagnetic properties of the original Infratonic but without the sound. This was much smaller and more convenient. She took it home on that last visit before Christmas.

During the next four months of intensive treatments, she made no further progress! We were all frustrated but nobody even considered that the change in the Infratonic transducer might be a significant factor. It was not until years later that I began to suspect the medical value of sound. In reviewing Shelley's record I acknowledged that the Infratonic had probably played a significant role in her initial recovery, and that I had probably thwarted her later progress by introducing a transducer without the stochastic mechanical stimulation.

Healing Through Coherence

Five years later, in 1994, I visited China's Army General Qigong Hospital, and was quite enthusiastic to discover that it had what may have been the most effective program in the world for treating paralysis. 50,000 paralysis patients had been treated at this hospital, with 90% showing some improvement and 46% showing complete recovery. Many of the patients had previously unsuccessfully undergone other treatment programs. Success after months or years of paralysis makes many of these recoveries especially extraordinary.

Wan Sujian, M.D. director of this hospital and our Qigong instructor, originally got the idea for this unique treatment facility when he was sent by the Army to a city in northern China to aid survivors of an earthquake that killed half a million people. About 320,000 were left disabled, 38,000 paralyzed. With his Qigong skills, Dr. Wan could help a few, but he was overwhelmed by the enormity of the disaster and saw the need for a systematic and effective method for treatment of paralysis. This was when he devised his plan for a hospital specializing in paralysis.

One by one, he started gathering children from across China who were gifted in Qigong skills, training them in therapeutic Qigong and putting them to work treating the paralysis patients who had been brought to him in growing numbers. When I visited he had 64 Qigong assistants and a 200 bed hospital.

The Qigong treatment methods practiced by Master Wan's students seemed complex, and were certainly unconventional by Western standards:

Tapping on the limbs with the fingers, using an irregular pattern of sharp strikes to break up blockages in the Qi and promote nerve growth.

Emitting Qi to supplement the patient's energy and activate the healing response.

Removing stagnant Qi from the body.

Emitting Qi to the hands and feet to reestablish the neural pathways to the brain.

Once a week, hospital personnel strap a patient to a framework that suspends him by the waist and chest in the middle of the room so that he is free to move his limbs more easily. Then eight Qigong doctors emit Qi to him in unison. Four stand in a square several yards away, maintaining and strengthening a Qi field, and the other four, standing closer together, work within this field, moving Qi within the patient and inducing movement. Master Wan explained that the Qigong doctors must establish a strong "Pei He," coherence with one another to be effective.

Pei He. I puzzled about that word. Like most Chinese words it has deep and complex meanings. It is not just agreement of cooperation. To have Pei He, a group must all be in accord, in harmony, be heading in the same direction and with the same intent or purpose. Members may be doing different things, but their individual activities must fit seamlessly in a single united effort. This, explained Master Wan, was an important part of why the eight Qigong doctors working together had a high success rate in treating paralysis.

The Chinese describe the Japanese during World War II as having Pei He, a tremendous group consciousness about conquest that made their country very powerful. Many Chinese believe that the first atomic bomb, dropped on Hiroshima was insufficient to disrupt the Japanese Pei He. It was the second bomb that finally dispersed the Japanese Pei He and helped bring an end to the war.

The Chinese see another incredible rise of Pei He on the other side of the world. Rather than a resolve toward conquest, the Americans have developed and sustained an incredible Pei He toward technology. Computers, jet planes, space shuttles, consumer goods. The same Pei He of American technology has penetrated the world and has rapidly industrialized China.

The Chinese have their own Pei He, an impenetrable attitude of peaceful cooperation that has allowed China to survive for thousands of years through countless invasions. With this powerful Pei He China absorbed and learned from its invaders but was never dominated by them. It maintained its own culture. The Chinese Pei He is a coherence of stability.

Qigong doctors insist that Pei He, coherence between doctor and patient is critical if treatment is to be effective. Perhaps this is similar to the Western concept of "bedside manner." Building this coherence is an important part of Qigong training.

Master Wan had his assistants demonstrate the exercise they used to develop this coherence. Outside on the grass, four students stood in the square about 15 feet across. Inside that square are another four students stood. I watched as the outer four stood in place waving their hands in rhythm while the inside four walked around in a circle making similar movements in unison. Everyone involved reported feeling the same Qi field. In addition all claimed they could feel the Qi strengthening within the circle.

I see a similar relationship while playing volleyball. I can feel unity when we are all working together. Everything runs smoothly and we often win the game. But if just one team member has a negative attitude, he can destroy the coherence. We would make mistakes and lose several points in a row. Is it possible that team play could be a form of Qigong training?

Etheric Template Transplant

4:00 AM

Years later, one of the doctors accompanying me on a Qigong training asked Master Wan if there were more to his paralysis treatment methods. This was an advanced Qigong workshop, so he was willing to share advanced secrets with us. He explained that the coherence within the spinal cord of the patient was extremely important. Deep trauma and injury could disrupt this coherence, and the spinal cord would essentially forget how to heal. Master Wan's secret method was to utilize his creative visualization, and take the energetic template of a healthy section of spinal cord copied from a healthy individual and essentially splice it into the energetic template of the patient through intention. This would provide a jumpstart where healing had stalled. While this was not sufficient in itself to overcome paralysis, it was a powerful tool in his toolbox that was an important part of his success.

It seems this technique is easily applied by anybody who has developed a clear, coherent, and solid skill in creative visualization, and of course, who believes it is possible. Further, it appears that this technique might illustrate a fundamental mechanism in the workings of prayer and creative intention whenever they are applied for healing. If this is true it would seem that every doctor, every medical school, every healing profession, should have extensive training in this creative visualization. It may be a huge part of why placebos are effective.

My friend, Dr. Jerry Alan Johnson, is a Qigong teacher with a rigorous Qigong training program. Before his students can graduate they must pass a challenging test. They are presented with a patient who has a tumor, a lump somewhere in their body. If the student is unable to dissolve that lump through Qigong healing, he is not ready to graduate. Master Wan's method involves inserting a new template. Master Johnson's method involves extracting or disrupting an unwanted template such that the lump has nothing holding it together and dissipates.

The Etheric Template

Why is it that oranges can sit in a bag for weeks, with every orange looking almost as fresh as when it was first picked? Then, an ugly green mold attacks one of the oranges. Within a day or two, the orange is covered with the ugly green stuff. Logically, it would seem that the oranges in direct contact with the moldy orange would immediately be infected as well. If you examine the oranges closest to the bad one, however, you find that they are unblighted. Rather, an orange or two on the other side of the bag may have started to grow mold. From the standpoint of Qigong, each orange has Qi, an organizing field or coherence that maintains its defenses against invaders like mold. Those with the strongest vitality remain highly resistant to the mold's attack for months after being picked. Those with weak vitality are defenseless against its attack.

The etheric template is a specific part of the vital field that surrounds the body. It is an energetic template that is in the exact pattern of the physical body, so that if any part of the body gets injured or damaged, the etheric template guides cells atoms and molecules as required to the exact place where needed and provides the exact plans to reconstruct the body. As long as the etheric template remains perfect our bodies heal rapidly and completely. However chronic pain and impeded healing are common. How is it, that one person can sustain an injury and be all better within a month, and another person can sustain a very similar injury, and suffer with chronic pain for 20 or 30 years?

The etheric template can be perfect and bring about perfect healing of the body, bringing it back to "as new" condition. However, it can also be warped by the invasion of Etheric Sludge: traumatic experiences that get stored in the tissue and limiting beliefs that are so deeply embedded in the body that we have no idea that they are there. When they become embedded in the etheric template, the plan of the etheric template, the blueprint for healing the physical body, becomes warped and we can have abnormal physical growth, impeded healing, or chronic pain. This 3AM writing develops the concept of embedded Etheric Sludge in the etheric template:

The life of emotional sludge is indefinite. It is important to dissolve it or otherwise neutralize it. Transmutation within the body field is a good method in that it is recycling. However, know that there is abundance, and it is not important to hang onto old energies. It is fine to pull them out and send them to the light. There's always more.

The emotional and mental debris that is lost during the clearings you do and orchestrate is fundamentally different from the stuff that leaks out during abdominal surgery. This is etheric Qi as opposed to astral, emotional sludge.

Think of gasoline in cooking oil. You can't put corn oil in the gas tank. Etheric Qi is made by plants, and to a small extent, people. Fresh plants mostly. This Qi is far from easy to cultivate. The Vital 1 comes close, providing fuel for many mental processes and encouraging the body to conserve its Qi.

So what is Qi? Qi is structured substance that is electric, magnetic, mental, and love from Earth and her plants and animals. It fuels active life, growth, and metabolic activity of plants and animals.

So...I still don't grasp what Qi is. Well, it's similar to the question to what DNA is. You might say DNA is an organization of atoms to make molecules. You might even specify which atoms in which percentage, but you're still not getting close to what DNA is. Similarly, within the world of the etheric substances, there is a particular substance which was designed and is ideal for fueling metabolic processes in humans. The body knows the difference between it and thoughts, emotions, and Etheric Sludge. As a matter of fact, San Pedro cactus causes the body to become acutely aware of the difference, tuning into Qi and pulling away awareness of emotional and mental stuff, pushing these residuals from the body. [*See Chapters 19 and 23 on San Pedro*]

You are right that this experience lets people discover the difference between Qi and emotional/mental stuff. You can't build an emotional world or a mental world out of Qi. Your world and the Earth are like oil and water; they just don't mix. They coexist, but do not interact.

Emotional and mental debris can sometimes displace some of the Qi in the body, like pouring sugar or canola oil into the gas tank. All are hydrocarbons, but they don't do the same job. A big part of the spontaneous healing of San Pedro cactus is the spontaneous release of astral and mental debris from the physical body. Astral and mental debris is never good for the physical health. It messes up the works. Spontaneous release causes spontaneous healing.

My 2nd San Pedro Ceremony and Head Colds

R.:   I had set my intention the night before the ceremony to embrace whatever changes and growth that San Pedro might offer me. Later that night I started feeling tension all up and down my right side, and particularly, a weird feeling of pain, numbness, and tension in my right jaw. This was hours before I drank the San Pedro. I had just called the spirit of San Pedro to prepare me for the ceremony!

I include this section here because it illustrates a very common form of Etheric Sludge that most of us feel frequently but don't consider to be separate from ourselves. This form of Etheric Sludge is thoughts, worries, and other mental residues that congest our brains and sinuses, and according to this, are the causes of many of our uncomfortable cold and flu symptoms. 12 hours after the ceremony, I wrote this:

Alma: The moon has now progressed across the sky and continues to hide behind clouds.

I applied the Scalene Light to try to reduce the uncomfortable feeling in my head. It's the same discomfort as last night except more focused on the sinuses. Blowing my nose helped, as did the Scalene Light. First I did Soothe, and then Clear, and it wasn't enough. So I tried Penetrate, and it seemed to help a lot. I thought I would write a little and then try it again.

Any suggestions on how to relieve the pressure? Excess thinking is compression. You are more acutely aware of it with San Pedro. You are given the opportunity to clear it out, principally by feeling it and relaxing to let go of the thoughts, the congestion.

The Scalene Light dissolves the excess thinking, making it more comfortable to let go of the Etheric Sludge and physical toxins which accumulate where excess thinking is congested. This helps where the pain is overwhelming. However, becoming aware of the pressure and letting it go through relaxing, etc. is a great teaching tool as well.

One interesting thing to discover is that colds and flu are methods by which the body clears excess thoughts from the head. If you don't clutter your head with trivia and worries, you don't get colds or flu. Or rather, you can still catch the virus, but you don't get the sinus congestion and discomfort that come with it. Allergies, similarly, are excess thoughts in the head. *This suggests that the Scalene Light is effective treatment for sinus related allergies.*

Good morning. Dawn has broken. Yet we have several hours before sunrise over the mountain. Roosters are declaring a new day. Catherine sleeps. *What shall I write about this morning?* Just write and see and feel what emerges.

Clearly the Scalene Light was helpful to cause my sinuses to clear, and most of the pressure in my head went away. I slept right away and stayed asleep until morning. This was a huge change from before the Scalene Light, when I was in pain as from a head cold and couldn't sleep. Colds are miserable. After the Scalene Light the cold symptoms were gone.

How much opportunity did I lose from using the Scalene Light? Once you got to sleep we were able to get to work removing the cobwebs. Suffering is not an answer. That's the old way, the Catholic way. Yes you are much better off with the Scalene Light. Stuff that's left over, Etheric Sludge, can stick around for a day or two, and reintegrate with your body afterwards. It is better to remove it with the Scalene Light than to have it remain in the body throughout the ceremony. When it is removed by the Scalene Light, more sludge can emerge for removal.

If you want to teach the "Power" of San Pedro, don't use the Scalene Light. If you want to provide maximum sludge removal, then apply the Scalene Light liberally after the first 8 hours of ceremony to remove all available sludge. The resulting clarity is amazing and no benefit is lost. For a rape victim, do the Scalene Light frequently the first time from the beginning. The second time, use the Scalene Light starting 8 hours after taking the San Pedro unless it is too uncomfortable during the early hours. More rapid recovery is achieved in this way.

I wrote this two mornings after I participated in this second San Pedro Ceremony:

I took a nap yesterday afternoon then slept for about 11 hours last night. Why the continued nasal discharge? You are detoxifying, eliminating limiting beliefs, thought forms from ages past. This requires lots of energy and sleep. Be patient with your body. It is working hard for you.

The San Pedro opened the door by bringing the buildup of Etheric Sludge to your conscious awareness. The Scalene Light dissolved the mental beliefs holding the astral Etheric Sludge in place. The clear water you have been drinking is lubricating, allowing the sludge to flow out of your body. The Mag 0_7 is assisting in this elimination as well. Your physical body is the principal focus of this purification, or rather, the etheric template which had been distorted with old emotional and mental sludge, is the focus. From our perspective the etheric template is the physical body. All pain and chronic disease are temporary manifestations in the physical body, created by distortions in the etheric template.

What else this morning? You forgot the Scalene Light before writing this morning. [*I got up and cleansed with the Scalene Light. Whenever I forget, I am reminded.*]

No birds. Nothing. Why so quiet this morning? Yes. It's Sunday night. Also, it's full moon. The moment of alignment. The world is pausing to breathe. Pause with it. Relax and breathe. Release everything you can. You are passing through a huge doorway, and initiation. Sleep again now and feel your body sinking into the full moon stillness.

IV. A New Perspective

21. Prayer

"There is no point in doing anything without full prayer and intent." is all about discovering what you want and either choosing to struggle for it or to receive it gracefully.

Prayer as a Foundation for an Intentional Life

I grew up as a scientist. I didn't pray. I didn't believe in prayer. To me things happened because of scientific principles. You apply technologies because they work. Cause-and-effect. You take an action and you expect the world to respond mechanically providing the outcome you anticipate. For me, if something doesn't make sense, I don't do it. I don't do things because people tell me to. I do things because I believe in them. I thought that prayer was something you did if you believed that you were not in charge of your life and that the God you believed in was.

Through all my adventures exploring the Chinese world of Qigong, I discovered that it was not simply a matter of applying engineering principles to cause the result that we want. It all got very complicated very quickly. A big problem was intention. These Qigong practitioners would intend that the invisible energies would flow up their legs or out their hands, then make parts of their body super hard or make steel super soft. Another big problem in my mind was that they were applying these technologies of intention to their own bodies. This is another violation

279

of the scientific method. You need to be an objective observer. You need to change just one variable and observe how it changes the result. If you go changing yourself, you create a huge problem regarding who the observer is, and the line between the observer and what is being observed.

A third big problem with Qigong is that it gets all messed up with the consciousness of living things. Qigong is not simply an application of technologies. It is a communication with nature, or in some cases an orchestration of the communication of nature. When a Qigong practitioner provides a treatment that is to accelerate healing, the practitioner is communicating with the intelligence of the body of the person being treated. Further, he is communicating with the Earth, the trees and the plant life in the atmosphere, so as to work with these entities so they can provide energy which flows through the practitioner into the patient. Often the practitioner will call in particular qualities from a healing plant, or as I mentioned earlier, call upon qualities of a disease state or intoxication with alcohol, to separate them from the patient's body.

Now we come to a fourth major obstacle to my understanding of the function of Qigong. Not only are plants intelligent and available to be communicated with, but the entire universe is available for communication as well. And it gets worse. Whereas we have physical bodies, we are, by nature, Spiritual beings. Our physical bodies appear to be separate from the rest of the world, to obey laws of cause-and-effect, to be able to discern the present, remember the past, and only guess at the future. On the other hand, our higher nature lives in an entirely different world, a world where past, present and future are one, where will is fulfilled instantaneously, where anything is possible. Further, this higher aspect of us is inseparable from the consciousness of the entire universe.

We were on the bus from Master Wan's training center to the Fragrant Hill Hotel northwest of Beijing. Master Wan was sitting in the front seat, and 52 doctors and therapists from the United States filled the seats behind. Suddenly, the driver slammed on the brakes and the bus came to a screeching halt just a few feet away from another bus. Master

Wan turned around to check to see that everybody was safe. He doesn't speak English and I don't speak Chinese, so if we are to communicate we must do so in sign language. I slapped my hands together to communicate that we could've had an accident. He paid no attention and just turned back to the front, and his student Little Zhao jumped up and waved her hands for me to stop. "No, no, Mr. Richard!!! You must not say that!!!!"

I thought about that lesson for a long time, and gradually came to realize that the message Little Zhao was communicating is that we are constantly creating our reality with everything we do and say. It was forbidden to speak a negative outcome to the Master, for our creative potential is truly powerful, and the Master is teaching us both how to express it more powerfully and how to control our expressions to create only good. Every thought, word and deed contributes to the outcome. If they are not coherent and focused on a clear objective, a powerful result cannot be expected and chaos is the most likely outcome. The Master simply did not receive or accept the possibility of a collision.

Withholding expression of a possible negative outcome, while important, is only the tip of the iceberg. We must learn to be very clear about our objectives and our purpose in everything we do. Only then can we think speak and act with coherent powerful intention. As you'll see below, I am being trained both to clarify my objectives and to increase the coherence of my thoughts, words, and deeds so that my work can proceed powerfully:

4:00 AM

Be not concerned with the details of life. As we said, we are taking care of them. They are the texture of the world you have created. Details are not of the planet. They are of your world. Nobody but you has your list of details. They disappear if ignored. As said, they are texture. You've created them to keep yourself busy, awake, and purposeful. Without them, your life goes on just fine. You have said to your office staff, and your brothers, "take care of the details." The result is that they take care of some details and the rest just fade away.

If details are the texture of your created world, what are its other features? Goals and objectives, possessions, friends and associates, enemies, and problems or obstacles. That's it. All are of your creation. None are features of Earth. Nobody else cares about them. They are created out of what you want or want to avoid, and what you believe.

If you change your wants and beliefs, your whole world changes; take this trip as an example: packing and preparing made up the texture of the last few days. As you get on the airplane, the whole texture of your life shifts. Vacation is so named because the details of your life are vacated, left behind. When you return home, a few details will have waited, but most will have disappeared.

The difference between what is real and what is important: Real stuff is part of Earth. Important stuff is what makes up your world. Unimportant stuff simply goes ignored, unnoticed. So you see, your world consists of what you notice and is flavored by what you assign importance to. Regardless of what you notice and assign importance to, we will hold you, guide you, and shape your life.

Prayer is a *HUGE* game changer: it doesn't change the Earth. It doesn't change your wants and beliefs. It simply reorganizes what you notice such that the details of your life are provided rather than struggled for. Do you understand? Each time you pray, you are consciously asking for what you want. We would provide what you are calling for anyway, but if you consciously pray for what you want, you will discover that it is all provided for you. This discovery that all your wants are provided for causes you to begin to study and observe, to organize and order your wants and beliefs. You discover that you ARE the creator of your life. You create the details, the worries, the fears. You create the joyful and happy moments. You create the abundance and prosperity, or lack thereof. You create the oneness or separation from the all. Details cause you to focus on lack and separation. Prayer causes you to focus on abundance, prosperity, and oneness.

Do you now see why prayer is such a game changer? It is similar to affirmations, intentions, to-do lists, business plans. It is all about discovering what you want and either choosing to struggle for it or to receive it gracefully.

If you believe you "don't deserve" you will find your heart's desires held from you. If you believe you need to struggle and work for whatever you want, then your life will be filled with struggle and work, and you will receive what you "want," struggle and work. If you pray and allow gracefully and without limiting beliefs, all of your struggles, punishments, withholdings, and need to work hard will fade away. You will find yourself moving from the architecture of your world, from the details, to the gentle colors of your world, to the Joy, Love, and Bliss.

Prayer works regardless of whether you pray consciously or unconsciously. Conscious prayer involves not only stating your desires clearly and verbally, but also (and this is of greatest importance) keeping track of your prayers and observing what emerges in your life. By observing the results of your prayers you quickly see the products of your unconscious limiting beliefs. You are able to identify these limiting beliefs, and are able to pull back your power from them.

Prayer is, then, not asking a higher power for what you want, but rather, a pathway to oneness, prosperity, and abundance.

Icebergs of Opportunity

For a long, long time you have searched for answers without ever knowing what the right questions were. Now, you have an inkling of an idea of what it's all about, but not much more than that. We've explained it to you over and over again, but you miss the essential ingredients. You have even taught it many times without grasping the key fundamental truths. We can tell you again, but it is not likely to make much difference.

The value of prayer is the tip of an iceberg, one example where you have known the key all your life, but have not applied it and have not realized how extensive it is; how it makes virtually all knowledge obsolete.

Not only can you create your own reality, but your reality doesn't exist. It is your mind, interpreting the Earth as the world. You still believe 99% of the illusions you have created for yourself through the ages. You could snap your fingers and all that would vanish, leaving you free to create at will.

The fact is that you are terrified of the prospect that you might dissolve all of your illusions and not be able to re-create them just as they were. It seems so much more "controlled" to change just one little thing at a time. Anything and everything is possible this instant and every instant. And yet that's still just the tip of one iceberg.

Another iceberg could start with the concept of dissolving limiting beliefs. You made big strides here yesterday by applying that which you were taught many years ago with Ellie. Through intention, you can breathe out stagnant electricity, not just isolated pain and inflammation, as you are attempting to apply the technique, but to all limiting beliefs, all at once. If you simply breathe them all out at once, what would happen? Would you lose purpose to your life? What would Etheric Sludge science be without Etheric Sludge to study and remove?

You identify with this task, not because it is true, but rather, because it is a mass belief which appears to imprison millions of people. Not only is liberation within your grasp, but as before, the only reason it is hanging around is because <u>you will it into existence during each successive in-breath.</u>

Another huge illusion is based around who you are relative to God. It used to be that either you didn't exist or God didn't exist. Now you discover that you can speak to God and God can speak back to you. Even more, you entertain the idea that God is within you, and that you can "pretend to be God" for a little while and "step in" to command the world to conform to your wants and desires.

Even the concepts that you can "be" God and God can "be you," hide the grandeur of the true reality. The act of smelling a rose is so hugely illustrative of all of this, and yet you don't see.

[I think of the way the shamans pray as they prepare the San Pedro cactus. They pray for it as it grows, as it ages, as they harvest it, as they cut it in preparation for cooking and as they cook it for many hours. They pray for ceremony participants to get great benefit from it throughout the ceremony. This would seem analogous to smelling a rose in that, as the shaman prays, he is instructing and guiding the San Pedro cactus in its purpose. Similarly, as we smell the rose and praise its beautiful scent we may be guiding it through positive reinforcement to cultivate a more and more beautiful fragrance.]

Another huge area is the field of will and intention. You have tracked down the will center to the Hara, and have come to believe that the will results from "concentration" of "attention." Again, these have been useful constructs through the ages. Yet they do so much to cloud the true and broad wisdom that can be derived from the simple exploration. Everything about you, or at least that part of you which you identify as you, is the product of a concentration of intention. You are, in essence, will – made – manifest, nothing more, nothing less. This is a huge realization, and one that is right around every corner waiting for the day, hour, or minute that you set down your illusions and fully embody your focused attention.

Again, fear is huge. What if all of your life were nothing more than concentrated intention or focused attention? What would happen if you really allowed yourself to study the substance and function of focused attention? Thousands of books are written on this topic, and yet you have chosen time and time again to read but not to apply.

Now the time is upon you that calls forth the wisdom of focused attention. It waits for you to pick it up, dust it off, and first, start to apply it gingerly, then recoil from its raw power, then to dabble some more. Through focused intention you have the ability to fully engage, not only in your little world, but in life on Earth. Learn this and you shall be ready for full terraforming. All that stands in your way is for you to know that this is so! Harness the magic you express every time you smell a flower or taste your breakfast, see a stoplight, listen to a bird, or enjoy the silence.

This is enough for this morning. You can know that it is all true in your next inhalation, or you can spend 1,000 more lifetimes skirting around the topic. The choice is yours and will continue to be right in front of your nose until you finally decide to choose it. Then follow through and choose it every instant.

Prayer and Awakening from Spiritual Slumber:
Words of Awakening for Laura

Laura is meant to awaken now. To do this she must decide with determination that it is to be so. She must pray that it is to be so. She must watch how her life changes. She must watch for the strange events, the anomalies. She must grant herself the Grace, the right to be perfect in an imperfect world, then to be perfect in a perfect world of her creation. Each "imperfection" will point her on her way, loving her, showing her the remaining Etheric Sludge within her from which she has the opportunity to pull back her power.

How can Laura keep her crown clear? A Scalene Light would be very helpful, but they are not ready. She can, every night before bed, and every morning after her shower, feel the Etheric Sludge in her head, heart and abdomen, and pull it out. As she pulls it out, she will feel it moving. After it is gone she will feel lighter.

Giving it to Archangel Michael is most practical as it cleanses the environment. Since Archangel Michael, along with other nonphysical presences on Earth, are prohibited by convention of free will to assist humanity unless asked specifically for intervention.

Laura can invite St. Germain to dissolve all Etheric Sludge within her and within her world. St. Germain will do this joyfully when asked. But Laura will need to ask regularly, at least twice a day, evening and morning. St. Germain can only act in the Now, which is for as long as the petition remains conscious in Laura's thoughts.

She can also call in Archangel Raphael for healing. St. Germain and his Violet flame can leave some rough edges, so it is best to ask Archangel Raphael to bring about deep healing and wholeness within Laura every night and every morning.

At first, Laura will find that she forgets or that she doesn't have time. With repetition, she will find that it gets easier and easier to remember and complete the task. 2 minutes with each will prove to be enough in the long term. However, 10 minutes with each at first will provide needed deep cleaning.

Appreciation:

A key part of any and all success is appreciation. Appreciation makes the experience appreciate in value for her in the future and for all who follow the pathway that she blazes.

She has the determination to summon her will and to rise to the opportunity, to persevere. In addition to appreciating Archangels Michael, Raphael, and St. Germain she must, most of all, appreciate herself for the incredible creator she always was, is, and has chosen to become. Yes, she has chosen to become. Laura has chosen the pathway of liberation from bondage of the illusions of Etheric Sludge, and has chosen to enter upon the lighted path of Christ Yeshua and the Magdalene, the pathway followed by all great sages and teachers throughout the ages.

Laura asks, "Who am I to be a great teacher or a great Sage?" We respond, "Who is she not to be? If she doesn't step up, who will? If she does, how many thousands will follow?" It is an easy choice. She and every other reader of this material chose this pathway before birth, and are called to choose this pathway consciously Now.

The importance of conscious appreciation, and acknowledgment that this is her pathway, cannot be underestimated. Affirming this pathway daily will help to embed it in the consciousness of her physical body. This will set the dynamics in place that will eject all conflicting limiting beliefs, which is *HUGE*.

It sets the stage for the "burning ground," the rapid-fire emergence of limiting beliefs into consciousness and "lifting them up to God" or "pulling back her power," which are the same thing from two perspectives. Realizing that they are illusion and calling upon a higher power to carry them away is the first step. Embodying that higher power and pulling back her Power from the illusions completes the process and leaves her standing "at the right hand of God."

"Who am I to believe, and claim, that I have the right to stand as a co-creator with God?" We respond, "Who are you to continue denying your birthright?"

The burning ground may involve a little meeting of resistance in your world. Church, government, family, and depressing and discouraging "friends" will emerge and expose your internal limiting beliefs. As you pull back your power from each limiting belief, you will find that the resistance in your outside world dissolves. However, the process initially feels like walking through fire.

As you step fully upon the path you will find that what was first an excruciating burning rapidly evolves into a sense of Joy, Love, and Bliss. Every painful new limiting belief you encounter, within you or within your world, becomes another exhilarating celebration along the path of reclaiming your birthright.

Every journey starts with a first step. You have already taken your first several steps, and you can feel that what previously felt as painful, has taken on an increasing flavor of joy.

Terraforming with Prayer

We have gone to work to deliver what you prayed for. Keep it up. Clarify. Embellish. Expand!

Also, get passionate about your prayer. To do this, ask for what you are passionate about. Speak each prayer and see how you feel. If you don't feel passionate change it a little or a lot until you are sure it is something you really want.

So you see the trick is not to say it in a passionate voice, but rather, to know clearly what you want and how much you really want it. This is difficult training for you. You have spent your whole life denying your wants. We have delivered everything you have been clear and passionate about, but that has not been a lot.

The everlasting now is all around you. Your attachment to past and future imprisons you, separates you from the Now. When you are lost in thought of how things were or how they should be, you are not present. Stay present.

San Pedro cactus can wait. You are doing good with the Vital 1, Infratonic S, and Scalene Light. Wait until June 20 for San Pedro. Savor the anticipation. Feel the anticipated bliss within your body. Know that you are cooking in your own way. You have, at last, started a life of prayer, of clarifying what you are passionate about, and of voicing it. This brings us great joy and relief. This clarification and voicing is such a big part of the picture.

The San Pedro cactus is extremely good at prayer. Over the millennia, it has clarified exactly what it wants with great determination. The result, as you know, is an intensity of etheric template which is unmatched by most plants. Each plant has its own deeply felt intention, which is expressed in not only its growth and survival patterns but also in its pharmacological effects. Each plant holds a gift for humanity and maintains it through fierce, intense prayer, constant repetition of desire.

In the case of San Pedro, humanity takes part. Your shamans throughout the ages have identified pharmacologically active herbs, and prepared them with loving ritual with lots of very clarified and focused intention. This is what has given these plants their very strong and diverse healing actions. A prayer and invocation goes out, not only to a single leaf or twig but to the entire species around the world and beyond, not only calling in the Overlighting Deva of the species for the intended healing, but also as a command to all plants of that species to become more effective at what they do. As a result, they intensify the concentration of effective alkaloids and fortify their etheric template with this amplified intention. When you take a medicinal plant, you are filling your etheric template with the essence of that intention, and for the time that it is active within you, you become the etheric template of that plant, leaving behind your humanness, and becoming plantness. With such a radically shifted etheric template, your disease processes are driven off.

This is a fundamental principle of shamanism. Repeated prayer and intention create a healing potential in nature. This blazes a new trail for others to follow.

As you can imagine, these intention-induced effects on the human etheric template are very easily influenced by the individual receiving therapy. An individual who does not want to believe, or who wants to "prove the shaman wrong" can, when imbued with the etheric template of the plant, seriously erode or modify the intent or effectiveness of the plant worldwide. This is why it is so important to respect the shaman and his work. When he or she brings you into his world, he is opening the door for you to be healed, and at the same time gives you an opportunity to do damage to the world of intention he has meticulously and lovingly nurtured and enhanced over the years.

Do you walk in a garden and appreciate its beauty? Or do you trample on the flowers. Appreciate the work of the shaman and you will empower him in his work. Always respect, appreciate, and say prayers for a shaman.

Shamanism is easier now because the old world is mostly dead and disintegrated. Abuse and condemnation are becoming less and less common, and loving appreciation is coming more and more. This is a time for shamans to flourish. You are a shaman, with your seed enhanced Vital 1, your Hydronizer, Ra-Ad, Mag Gen, subspace transportation and communication. This is all about teaching nature new tricks, or in most cases, reviving old tricks and modifying them through prayer and intention, ritual and repetition, to forge new laws into the world of nature. Know that, as an inventor/creator, you are not so much discovering the laws of nature as rewriting them.

Thus, you will now see how important it is that you pray, intend, and invoke every morning regarding what you want. Just as the San Pedro cactus is programmed and potentized through endless repetition, your life is also programmed by your heartfelt prayer and intention. It is not so much that you are discovering what you want, but rather, you are creating, potentizing, shaping your desires every minute, and in particular, every morning when you start a new day fresh.

To some extent astrological patterns and the Will of God are guiding your destiny, but to a much larger extent, you, through your Morning Prayer and invocation, are shaping, magnetizing your life. Your prayer today, combined with your prayers of the last 10, 100, 1,000 days, shape your causal etheric template, creating a powerful force that brings your life experiences and creations into being.

Pray and invoke daily with great intensity and resolve, and in 10 days, you will see a whole new world emerging. The first few days are the most noticeable because you have expressed so little resolve in the past. Once you get beyond 10 to 20 days, you are building upon an edifice of similar intention so don't notice as much of the effect, but the result is cumulative. Your life will become more and more galvanized, magnetic, drawing to it those events and creations that you have repeatedly prayed for and invoked through heartfelt ritual.

Go forth and rewrite your future!
Revel in the Ride and Embrace the Abundance.

22. Intention

"Intention stands outside of the art of science
as it is currently practiced."

4:00 AM

Intention is poorly understood, principally because it is not a product of the mind. How can we understand something that is outside the realm of understanding? The mind understands the law of cause-and-effect. It grasps the concept of time and how things must be created step-by-step over time.

Intention is a creative act. The act of intending sets the entire universe in action fulfilling the object of intention. The universe is not limited by cause-and-effect nor is it held back by past, present and future. You can set your intention now, and its fulfillment begins in the past. It is as if the universe anticipates your intentions and starts long in the past setting the wheels in action to bring about its fulfillment. From our perspective it all happens now. They manifest instantly and any changes in the past are the result of the current manifestation. The products of intention emerge spontaneously, serendipitously. You might say magically.

If there is a physical organ of intention it is the Hara, or lower Dan Tian. It is located something like an inch below the navel in your lower abdomen. That which is placed in this place, this home of the will comes into being in your life.

Every time you focus your intention solidly and fixedly on a particular thing, a thought, want, or desire, you are exercising your will and becoming more capable of holding an intention. While it is a valuable exercise to hold the thought of intention in your conscious mental awareness over a period of time, the actual creative act is placing this intention in the will center.

Knowing that it will be done is a powerful active will. Doubt is a powerful act of will as well. You can prevent almost anything from manifesting by placing doubt in your creative center. Any other belief you have, which is a solid part of your belief system, is acting from the standpoint of will, and is creating your world every minute. This is why limiting beliefs are so powerful, particularly unconscious beliefs like doubt and shame and "I don't deserve." We have no idea that they are controlling what we intend. Or rather, they are shaping, coloring, blocking, or defeating whatever we believe we are consciously intending.

If you believe in Murphy's Law: "Whatever can go wrong will go wrong," your intention, or at least your hidden intention, is attempting to shape all of your intentions to conform to this belief.

Releasing War and Disharmony

[I wrote this a week before 12/21/12]

Today's discussion is on peace and harmony. These are principal characteristics of the New World. They really are one thing.

First, there is war and disharmony within you and within most humans on Earth at this time. Releasing this is an objective over the next 2 ½ weeks, within yourself. As you can release it within yourself, you will release substantial amounts of it in humanity, which is just a byproduct, a reflection.

How do you release the war and disharmony? First, do not get involved in conflicts. Sending legal papers seems to be a warlike, disharmonious action. In fact, it is the way of the world. It is a method to resolve conflict, to avoid war through the application of consistent rules. By submitting the dispute then letting go, you are able to maintain peace and harmony within yourself, which is your objective. Appreciate the legal system for its ability to allow you to purchase peace and harmony. The key is, as always, choosing and letting go.

Use the teachings of "Busting Loose" to pull back your peace and harmony from the conflict and from the legal system. This is extremely important. If well done, it will lead to the collapse of war and disharmony within the legal system, along with the collapse of the legal system, which also must collapse when there is no need for it.

Know that this and every other event in your life is purposeful, effective, efficient, and strategically planned by your Higher Self, "The Teacher," to be the best way to get to readiness. Apply your will and your intention clearly and strongly with great power. It is this direction: Apply every opportunity to pull back your power from war and disharmony, wherever you perceive it, focusing on feelings within yourself.

You will find the roots of the world's war and disharmony within you, these pockets of war and disharmony within you are buried deeply inside of you. All forms of inflammation disorder within you are the root cause. Become 10 times, 100 times, 1,000 times as sensitive to what is going on inside of you and you will discover these buried seeds of war and disharmony.

This is a component of the big shift. It is what allows the advanced technologies to emerge on Earth. Without peace and harmony, these technologies are not safe in the universe.

Be on the lookout for all vestiges of war and disharmony in your interactions, search for the source within you. Do lots of quiet meditation, feeling deeper and deeper into your body. This will require a high degree of focus and attention. Its value cannot be underestimated.

Doubt - You doubt all this is possible. Find the doubt. Dig high and low, in and around your body. Find the doubt and call back your power from it. Doubt is only your amazing creation, something you put in place to hold yourself in place. Pull back your power from the doubt and command it is so.

What is Your Intention?

Be strong and clear in your intention. You have one week. Set aside all else, and focus only on this. That's right! This is so important that we want you to set aside all distractions and to move forward with this. Include in your intention that all your physical world tasks be completed perfectly by those who are doing them, including you.

You will find that you can split. You can remain focused on your creative intention as your body goes about its daily routine. This seems confusing but it is not. You must disengage from your physical body and merge with your spiritual essence so you will be ready to return completely and fully to your physical body on your birthday.

Abstract more from the physical, becoming more and more aware of yourself being spread across the entire universe. Focused attention needs to be reshaped into focused creative intention. On your return, your life will be based around creative intention rather than focused attention. Embody your intention rather than your attention.

My Creative Intentions:

1. I let go of all limiting beliefs, emotions, traumas, expectations, etc. and replace them with creative intention.
2. My life and the lives of all of humanity are based on peace, joy, harmony, and abundance.
3. My creations and inventions fall together effortlessly.
4. My business accelerates effortlessly.
5. All my possessions sort themselves out so they are simple and effortless, requiring no significant effort on my part.
6. My health, diet, and circle of friends, associates and acquaintances become a complete source of joy and bliss for me and all of humanity.

Exercise to do Every Morning:

Every Morning: Be specific about a goal setting, commanding, and prayer. Then observe the unexpected, and when you don't feel comfortable or feel the tugging or tightness of your womb space, pull back your power and make new choices.

Seed the uncertainty with new, innovative choices. Do things such that you don't know the outcome. Appreciate, and make new connections. Notice the unexpected and you will be a person around whom innovative things happen.

Be loving, appreciative, truthful, honest, commanding, and receptive. Expand into your new role as a world leader. Pray and let go and your world will be filled with unexpected abundance and innovation. Be patient, accepting, and appreciative of yourself and how you have been. Follow your inner compass, and choose to pursue the acceleration of your life.

"There is no point in doing anything without full prayer and intent."

23. San Pedro, Appreciation, and Intention

Vilcabamba Theories of Longevity

I learned after I purchased my property in Vilcabamba that Vilcabamba is world famous for longevity. It had one of the highest per capita centenarian rates in the world, which means the highest percentage of people living to 100 years old or more. My land was owned by Miguel Carpio who lived to be 127 years old. The house where he lived still stands. Thus I was quite intrigued by the concept of longevity, and what factors might contribute to such long life in this community.

It turns out that many researchers, mostly in the 1970's, approached this question enthusiastically and with a variety of perspectives. The one that seemed to get the most traction is the notion that the soil in Vilcabamba contains an abundance of minerals so that people who eat the food and drink the water have complete mineralization and thus live longer. This may be true. However there are lots of communities with similar mineralization. Also, I remain skeptical of this view because this perspective supports a huge nutritional supplement industry. Where there is profit in popularizing a perspective, that perspective will be popularized enthusiastically and relentlessly, regardless of truth.

The second theory holds that the relaxed lifestyle and year-around friendly weather account for the extraordinarily long lives of the people of Vilcabamba. It is true that I felt relaxed and comfortable as soon as I arrived, and this fact is prominent in my looking forward to returning there. In addition, books written in the 70's described the diet eaten by the centenarians. They ate no processed foods, and often ate meat no more than once a month. Certainly, Vilcabamba is a Garden of Eden. I found on my land fruit trees of a wide variety, including avocados, cherimoya, coffee, orange, banana, and many others. Green vegetables of all sorts grow spectacularly well. 10 or 20 miles to the east is the Amazon rain forest where it is frequently cloudy and gets lots of rain. 20 miles to the west is much drier and hotter. Also, there is a great deal of medical research, which shows the high-value of such a diet to longevity. However, there are many communities nearby that would seem similar to Vilcabamba, which did not have the high rate of centenarians.

One interesting bit of research presented in these books about Vilcabamba discussed the role of vitamin E. The first study showed that cells tend to multiply about 7 times before dying out. It showed that this pattern happened in a wide variety of circumstances. This was used to argue that lifetime is limited and aging is inevitable. Further it would seem to explain why animal studies which restricted caloric intake to 50 or 60% of what the animals would normally eat, slowing the metabolism, brought about a profound increase in longevity.

Another book contradicted this first book, bringing up the role of vitamin E. It acknowledged that cells tend to multiply about 7 times before dying out, but pointed out, that if the cells were provided with an increased supply of vitamin E, this multiplication rate in the lifetime of these cell cultures would be up to 5 times longer, essentially being able to multiply 35 times. This is very interesting in itself. It suggests that one of the components of longevity is being sure you get an adequate supply of vitamin E. Not only is it a good antioxidant, but it contributes to cell longevity and fertility. The falling fertility rates around the world suggest that a predominant vitamin E deficiency might be one of the cofactors of this trend. All the same, I am aware of no research to show that the Vilcabamba residents got a higher percentage of vitamin E in their diet than other nearby communities.

In my research, I noticed that one of the outstanding qualities of life in Vilcabamba was in the area of cardiac health. Vilcabamba residents had much healthier hearts than those in nearby communities, with reduced blood pressure and reduced incidence of cardiac disease. People would travel to Vilcabamba with heart conditions and find that their heart problems disappeared. I searched for causative factors for this improved cardiac health, and the only thing I could find was that San Pedro cactus, a cactus somewhat like peyote cactus, which grows naturally in the western Andes of Peru and Ecuador, and is used by local shamans as what they consider to be powerful medicine, proved to be responsible for substantial improvements in cardiac health.

It appears that each continent has its psychedelic medicine for clearing limiting beliefs. Across the border from my home, in Mexico, "Ebogaine treatments" are offered for drug addiction. Ebogaine is a drug extracted from the bark of the Eboga tree in Africa. European mistletoe is another medicinal plant used by shamans for thousands of years. Whereas San Pedro is used on the west side of the Andes in South America, Ayahuasca is used on the east side and Eboga is used in Africa for a similar purpose. There are probably many other such plants in other specific locations which are used for clearing limiting beliefs.

Could it be that regular use of San Pedro cactus, a psychedelic drug, could be responsible for the improved longevity of centenarians in Vilcabamba? Seems like an outrageous theory. I asked a local young shaman resident of Vilcabamba. He insisted that yes, use of San Pedro cactus was the secret of longevity for the centenarians. He said that the tradition was for them to use it every 3 to 4 months during a full moon to purify their bodies and extend their lives. San Pedro cactus grows throughout Vilcabamba, and I saw a forest containing thousands upon thousands of 10 and 20 foot high San Pedro cacti not far from the main road. Also, the nearest neighboring town, just around the corner, is named San Pedro de Vilcabamba, which I am told, has several forests of San Pedro cactus. This it appears that the centenarians of Vilcabamba had the means and the motivation to use San Pedro cactus on a regular basis.

Countering this was the testimony of my taxi driver, the grandnephew or great grandnephew of Miguel Carpio, the centenarian who lived to 127 years old on my property. I asked him what he thought of the theory that San Pedro cactus was responsible for the longevity of the centenarians. He stated categorically that this was false, that San Pedro cactus was virtually unknown more than 20 years ago, that it was very unpopular to the local residents, and only used recently by a handful of visiting tourists. To illustrate his point, he described a man who got high on a drug and murdered his wife. He wasn't sure whether the drug was San Pedro cactus or marijuana. I asked if he knew the secret of the centenarians' long life. He answered that he thought it was the diet, an extremely limited and, to him, undesirable diet. He admitted that he and his uncle had both recently had surgeries for diet related conditions, but insisted that he would rather live a good life, enjoying good food, than to live a long life on an undesirable diet.

In the book, "The Centenarians of the Andes," David Davies travels to towns around Vilcabamba, and finds that to the north and to the south, there are several towns with centenarians. All are on the west side of the Andes, just over the hill from the Amazon rain forest. All have a town named San Pedro nearby. Vilcabamba has "San Pedro de Vilcabamba" 6 miles north, which has several large groves of San Pedro cactus. Also the town with the highest concentration of centenarians, and with two of the oldest centenarians he had found, was named "San Pedro de la Bendita." Further, Vilcabamba used to be named Wilco Bamba, after the Wilco tree, the bark of which is used to make another psychedelic medicine. There used to be lots of Wilco trees in Vilcabamba and they are still highly praised for the good fortune they bestow on the land and its occupants. Wilco and San Pedro are often prepared together. In more recent years, the residents of Vilcabamba have cut down most of the trees for firewood, including many of the prized Wilco trees. Such is progress.

The jury is out on what contributed to the extraordinary longevity of Vilcabamba centenarians. San Pedro Cactus seems to be the only factor that might set Vilcabamba apart from other communities around the world. Even here, we have confounding variables. San Pedro cactus grows well in much of the southern and western United States and is a very popular, drought resistant plant, wherever water supplies are limited. Take a trip to your local Home Depot hardware store, and you are likely to find San Pedro cactus for sale among the cacti.

The simplest way to find the answer to the longevity mystery would be to ask the centenarians what their secret was. Did they even know what their secret was? Unfortunately, the invasion of the longevity researchers into Vilcabamba implanted a variety of factors, which undermined the longevity of the centenarians. With fame, came free dinners, along with increased caloric intake and lots of meat. With the arrival of Western medicine, came antibiotics, which killed off intestinal flora and impaired the digestive system. Some hypothesize that antibiotics are what brought an end to most centenarians in Vilcabamba. Then there's the theory that the centenarians simply didn't know any better. They were revered for their longevity, honored as important members of the community. The Western doctors were incredulous that anybody could live to be well over 100 years old and still be working in the field, reading without glasses, having a healthy appetite for sex, and taking care of themselves. They basically conveyed the idea that, "you should be dead by now." Many researchers feel that the "placebo effect" is so powerful that the opinions of the invading doctors and researchers are what killed off the centenarians.

In any case, the centenarians are now gone. There is nobody left in authority to ask whether their secret was a hallucinogenic cactus which, by the way, was condemned by the Catholic Church, a dominant opinion leader in the community. The centenarians have passed away, but an insatiable desire to know their secret lives on within my heart.

So, in this chapter we explore the shamanic medicine, San Pedro cactus. The literature tells me that San Pedro cactus is good to reduce blood pressure and improve cardiac health. In addition there are many claims that it brings about a wide variety of seemingly remarkable healings. If this is true, why is it so ignored by the modern medical profession and so reviled by the church, by the local community in Vilcabamba, and apparently by the United States government which allows people to grow San Pedro cactus in their gardens, but not to prepare it is a medicine?

If you are looking for a powerful hallucinogenic and a fun ride, there are far better choices than San Pedro. As far as I can tell, psychedelic doses make you sick and uncomfortable. If you want to become open and grounded, a San Pedro ceremony with lots of prayer and intention and not much San Pedro is likely to provide spectacular results. If you're not interested in any of that but are interested in how the Scalene Light, the Infratonic S, and the Vital 1 work, then you will absolutely love this chapter. Well, that is if you can handle the emphasis on prayer, intention, the intelligence of the Devic kingdom, and how Community (with a Capital C) involves so much more than getting together at Starbucks.

Let's start with a little background from:

What are the Medical Uses of San Pedro Cactus?

"In Andean traditional medicine, San Pedro cactus, or *Echinopsis pachanoi or Trichocereus Pachanoi*, was one of many medicinal plants believed to cure unknown illnesses. Research has shown that preparations of the cactus may be beneficial for the circulatory system, as they lower high blood pressure and reduce the risk of cardiac disease. The plant has been found to treat nervous conditions and alleviate joint problems. San Pedro cactus possesses powerful antimicrobial properties as well, preventing the growth of more than a dozen strains of penicillin-resistant bacteria.

"Although San Pedro cactus is generally praised for its antimicrobial properties, specifically its ability to inhibit multiple strains of Staphylococcus, the plant is met with much controversy. The cactus contains high concentrations of mescaline, the same psychoactive compound found in peyote. Traditional preparations of San Pedro cactus were originally used to induce meaningful visions for Andean tribes, a practice now performed by individuals seeking to experience a drug-induced high.

"This has created a gray area regarding the cultivation of San Pedro cactus. Although it is legal to grow the plant for ornamental purposes, it is considered highly illegal to grow it for consumption."

I have taken San Pedro cactus twice now, 3 months apart on the full moon. The first dose was a very small amount which seemed to serve the purpose of putting me more in touch with Nature, though I didn't have much of a psychedelic experience. Even with my 2nd dose, somewhat larger, I wouldn't describe my experience as psychedelic. I got no visual images, but rather, a connection with Nature, and a knowing, as if Nature was speaking to me. I felt like I was dancing, and weeks later I still felt a gentle sense of dancing within me. For those seeking psychedelic experiences I suspect that LSD, Psilocybin, or DMT might provide what they're looking for. San Pedro cactus, as I experienced it, seems good for little more than getting connected with nature and having a broader concept of community.

As we go further, I am taught that the consciousness of San Pedro, of all the San Pedro cactus in the world, is a single being. The role of the shaman is to communicate with this Great Being of the San Pedro cactus and to request that it bestow its healing on ceremony participants.

First, Some Background

There is a wonderful book on San Pedro cactus, which adds greatly to its understanding. It is "The Hummingbird's Journey to God", by Ross Heaven. If you like this brief summary, get the book. You will love it.

Ancient cave art shows that San Pedro ceremonies have been around for thousands of years. Then Catholicism drove San Pedro ceremonies underground. The San Pedro ceremony that emerged, and is still mostly practiced in Ecuador and Peru, is steeped in Catholicism and carries a strong flavor of suffering, making the commonly found ceremony quite uncomfortable. This is described in detail on pages 59 through 64 of the Hummingbird book.

San Pedro is considered able to open a doorway into a world beyond illusion so that healing and visions can flow from the spiritual to the physical dimensions. I view the world of illusion from which San Pedro offers relief, to be what I have called Etheric Sludge, limiting beliefs that buildup and cloud our vision so that we perceive what we want or don't want rather than what's really there. My impression is that the San Pedro ceremony is a very powerful medicine for removing Etheric Sludge.

The outer waxy coating and inner white pulp of the cactus cause vomiting, so it is claimed that that through these parts, the spiritual and physical toxins are removed from the body, almost the same as if someone were going through what is often called "exorcism." I have found that careful preparation of the cactus substantially reduces vomiting and nausea, but that vomiting still occasionally occurs, and when it does, generally proves to be very therapeutic.

In addition to mescaline, an "active ingredient" in peyote, San Pedro contains many chemical compounds. Some of these compounds mimic the effects of adrenaline and noradrenalin, the so-called fight or flight chemicals that are released naturally in our bodies to prepare us for action when reality shifts and we feel uncertain or anxious. Perhaps it is these as much as mescaline which commonly give us a sense of awe, and the awareness that we are in the presence of something mysterious and more powerful than ourselves. My impression is that a great deal of the Etheric Sludge which crowds our consciousness and creates illusion is anchored in the sympathetic nervous system, the fight or flight response. Thus these chemical compounds which influence the sympathetic nervous system might be key in the clearing of Etheric Sludge.

Interestingly, ayahuasca, another powerful hallucinogenic shamanic medicine, which grows in the Amazon rain forest on the east side of the Andes, tends to sweep us away from ordinary reality and into an abstract Spiritual world. San Pedro, on the other hand, which grows on the west side of the Andes in Ecuador and Peru, brings us closer to the consciousness of this world and exposes its beauty to us, which can be highly transformative, and even life-changing. It teaches us how to live in this world, the one that we know, but in balance and harmony, to release our old illusions and see the world through new eyes.

San Pedro is not a hallucinogen like ayahuasca or marijuana, so people will not see images and pictures. San Pedro's teaching is visionary instead, bringing us to an awareness of the natural world and our place in it. Rather than decreasing our awareness of the natural world like alcohol and anti-depressants, San Pedro clarifies our awareness that we live in a world of choice. It shows us that we are more powerful than society allows us to believe. Power can be frightening. It can also be liberating!

When we drink San Pedro, we realize that life and beauty are all around us. From this knowing we learn to act responsibly in a world based on nature, rather than a synthetic world based around technology.

The indigenous shamanic view is that all disease comes from our minds and souls. They don't believe in a purely "physical" illness. This is of course in direct conflict with modern medicine. Modern doctors are called physicians because they deal only with the physical, and attribute all other factors to psychosomatic beliefs, essentially illusions. Actually, these two beliefs are not that far apart. They both agree that there is a physical component and a mental/emotional component to illness, but the shamanic view is that the physical component is unimportant whereas the physician's view is that the mental/emotional component is unimportant.

In scientific cultures we are obsessed with a need to know, to understand. I suffer from this frequently. In fact this need to know is not a need, but a belief system. We believe we need to understand so we can explain away those phenomena in the world that don't make sense to us or conflict with what we want to see or are told we should see. Thus this need to understand is a big part of the disease itself. Those conditions labeled chronic fatigue and fibromyalgia appear to be caused by an addiction to this "need to understand." I suspect that the San Pedro ceremony might provide spectacular results for people claiming the labels chronic fatigue and fibromyalgia.

From here, we move again into some of the material I have written at 4:00 in the morning.

The Function of San Pedro Cactus

4:00 AM

This is a dark morning. The full moon has passed, yet the full moon energy is still present. *What do I want to learn this morning?* You're too sleepy to think. If you can keep writing and write rather quickly we can transmit some fascinating and difficult material.

The idea that San Pedro is similar to a device will be offensive to most, but will be interesting to a few intrepid scientists, partly because breaking into pieces and studying abstract parts of a truth remains the way of academia. It is true that this is one of the mechanisms of San

Pedro, and that it is also a valuable future healing modality. Researchers may be able to test and prove this for you. They face many obstacles, not the least of which is to convince the Research Boards to get behind this. The mechanism you describe is an important part of the mechanism, but is inseparable from the infusion of the body with a loud receiving antenna attuned to the actual San Pedro cactus.

How does San Pedro work? There are so many ways to understand it depending on the aspect of the healing effect that is to be emphasized.

Components:

1) Enhancing the nervous system.

2) Enhancing the etheric template.

3) Repressing the thoughts and emotions.

4) Opening the pineal to increase visions. (While DMT is increased, it is a very different process from that caused by ingestion of DMT.)

5) Being infused by the etheric template or Overlighting Deva of the San Pedro cactus. This Being is strong in your land and wherever San Pedro grows and is used. It is less strong where San Pedro is only brewed, and almost nonexistent where San Pedro is brought onto a land in powdered form.

The San Pedro Deva is, to a large extent, all of Nature. Yes it is somewhat differentiated to be more cactus in nature than other forms of plants, and it is more plant than animal or mineral, but it is really all of these things, including Gaia. A key part of the San Pedro ceremony is inviting in the San Pedro Deva to the brewing, making this inviting a ceremony in itself, preparing the brew so it carries the intelligence and the essence of this being. To a large extent, the cooking, or intention while preparing San Pedro cactus can draw out or enhance whatever aspects of the San Pedro intelligence that are desired. To learn how to prepare the San Pedro medicine one must clearly define what the intention is, then command it and ask. Different intents will produce very different results.

Intention and Appreciation

Storing it for months in a dark place horizontally, and regularly praying and speaking intentions, and appreciating the role of San Pedro in bringing community to Earth will allow the cactus to manufacture an alkaloid content which is ideal for the intended ceremony, preparation, and cooking method.

The San Pedro cactus loves its job. Like anyone else, it loves appreciation. Store it in a loving, dry, quiet environment away from radio waves and 60 Hz power.

Letting the San Pedro know its role through appreciation during growing will enhance the ceremony as well. All San Pedro cacti on the land will participate in the ceremony. Including the Spirit of San Pedro cactus on the land in the prayer process will make the ceremony much more powerful and healing.

Similar to an ocean of sea water, surrounding the participants in a sea of San Pedro energy during their preparation, and praying and singing for them, will help make it more potent and prepare them for the process. It is just as important to prepare the people as the San Pedro cactus, though is not necessary to freeze or boil them.

The San Pedro Ceremony on the Land

The shadow of the shadow is revealed with San Pedro ceremony. It pulls us from our emotional worlds and mental worlds into a physical world where we are one with nature, and into an emotional world where we are one with the love, the community of nature.

The day after the ceremony we are in a world of full grounding within the world of nature, with our emotional and mental worlds inaccessible to us. Through our nervous system we feel the heaviness, density, and sometimes pain of this heavy energy that is stuck in our bodies because of our residual thoughts and emotions. Yet we are unaware of their content, or rather, we are separate from their content. If we try, we can feel the emotion that is stuck, but we are aware that it's not ours, and we find ourselves willing and eager to be free of it.

The day after the ceremony offers a huge door to leave behind all this baggage we of held in our body, which is held in place as diseases, pain, and conflict in the world around us. If we leave behind all this emotional and mental sludge at the ceremony site, we are free, on the following days, to walk without it, or to recall it into our thoughts and emotions and rebuild it into our physical bodies. This is our choice. It is easy to stand free of it while in the Vilcabamba, in Ecuador, but as we return home to our old lives we feel the density of the Etheric Sludge we left behind. We experience it as oppressive, heavy, and even painful. We can choose to fight it or to surrender to it, as it begins to sink back into our physical bodies and consciousness as our old emotional and mental lives become "real" again. *[Note: it is most practical to clear your office and your home with the Scalene Light when you return from a San Pedro ceremony. You will find your world much more comfortable to live in without the residual, encrusted Etheric Sludge.]*

The day after the ceremony is the key and real awakening. We can let go of the concept of "recovery," and instead embrace the concept of "liberation." Constant movement such as, walking, yoga, and exercise, can help to free up the sludge. Introspection can help us to quantify the sludge as we stand separate from it. Extraction technologies like using the hands to pull out bad energy or psychic surgery can help, but it is our letting go which is important, not the pulling out done by another. Practitioners can temporarily reduce or remove it. Through our belief, intention, and expectation we either remain free of it or retrieve it, making it part of our bodies again.

Participants can be taught techniques of extraction and dissolution, re-absorption, and refilling the etheric template and body with light. They can work on themselves, but can focus much more effectively if they practice with a partner. Partners will help participants to remain focused and explore the hidden pockets of Etheric Sludge within them, encapsulating it with feeling and visualization, preparing it to be released from the body. Then extraction techniques will be useful, since the individual has decided to prepare and release the sludge. Sweeping, etc., will aid in the smooth, comfortable release.

Pulling back your power is a most effective technique which prepares valuable emotional and mental raw material to be reabsorbed by the body.

Similar to ginseng, maca, and astragalus, which introduce external substance to the body to boost vitality, clearing the Etheric Sludge in and around the body releases etheric Qi which was formerly bound up in congestion, much as erasing computer files makes more space on hard drives or in the memory of computers so they can work faster and more efficiently.

People are more energized and emotionally and mentally clear on the day after the ceremony because of the amount of clear, available emotional and mental substance in their bodies.

The Gift of San Pedro [My Morning after my first San Pedro Ceremony]

A spectacular morning in Ecuador. What do I want or need to know this morning? Life is not as it appears to you. There is no need for limits. All is made up of choices. Catherine is open to all of life. She speaks freely, openly, in the moment.

Anything more on the land? Don't assume you need to figure out anything. All is perfect! As you speak, it will happen. If you build it, they will come. Not "to a community" but on their own individual roads, toward finding a place in community. They want to serve, to create anew. Those who don't will stay away, or pass through without stopping. You need a place to shower, cook, and pee in the Carpio house, along with a future Museum to honor the centenarians and their longevity. Visitors can sleep in tents. If they want to buy propane they can have hot water.

Catherine must be strong, forceful, regarding staying and going. A free or cheap place to stay and an opportunity for community ceremony is why they come to Vilcabamba. Go light on the hallucinogenics. Use them only on the new and full moon. As the energy of the land grows, less and less San Pedro will serve. The integration of people, San Pedro, and the land will permeate the space. This is true community. People will be able to connect with the land, being called through the heart to serve. This energy will permeate the people, the water, and the land. It will flow down the stream, down the river, into the watershed, into the land. It will spread by permeating into the visitors. Then as they travel to their homes, their feet will root into their homeland in constant connection with Mother Earth. This can be enhanced through basic yoga or Qigong grounding; sun, moon salutation through the heart, connecting to the column of spirit while connecting to Nature through trees, plants, and the natural surroundings. All can be initially done during the peak of San Pedro, then the next morning, as people are coming down. Qigong practices may be used in connecting, in moving Qi, in healing through clearing, and in connecting with the Universal Life Force.

Methods for powdered San Pedro: San Pedro doesn't like being in powdered form. Grounding and opening are important keys. Cooking is a process of hours of preparation in love, structuring the water molecules with San Pedro. *How do you export the San Pedro cactus?* The plant doesn't need or want to be exported. It is a holographic process. Allow the "community" energy to potentize Earth. One ceremony, conducted with full heart and full intent, is felt around the world.

The purification before San Pedro is a process of singing, drumming, bathing, dance, prayer, intention, clearing surface sludge, and grounding in openness and union. It is our intention to become transparent in the process. Drink a lot of warm lemon water. This makes the water alkalizing, which is better for prayer and intention. With each glass, fill the cup with Intentions.

Purification after San Pedro

Find tension, focus on it, and dissipate it. Harmonize all your tensions. Be aware of how the body has become more physical, yet less mental. What is the nervous system doing? Do you have a high level of alertness? Utilize the alertness to study the consciousness of the body. Focus on physical activity and Earthly interaction. It is important to speak prayers and intentions through this process because the process dissolves all intention, and if you want to be focused on getting the most out of the San Pedro ceremony saying prayers for what you want resets your intention. [*This is true with use of the Scalene Light as well.*]

Use the energy and feel the sense of activity without competitive objective, without time awareness, without future, and past awareness. Focus on nothing but your connection with the land. Spend the day after the San Pedro cactus ceremony to connect to and work with the plants and animals. Feel the love in all things throughout the day following the ceremony.

The day after San Pedro is not a recovery; it is an integration. Notice your relationship with food. Feel the love in food. Eat small nibbles of food. Feel how it interacts with and modifies the function of the body. Wait a while and eat a little more.

Freedom from frustration: Notice that you are separated from the frustration in your body. The frustration, stress, and anger are there but you perceive them as not you. This is a perfect time to let go of it all. This is a big part of the healing of San Pedro. We let go of disease and illnesses we have been holding onto because we believed they were us. This is a perfect time to release it all now.

The day after the Ceremony, Richard has moved back into the attunement with San Pedro while writing this. You can call it integrative consciousness. Like the Scalene Light, San Pedro teaches a disengagement from past and future to present, from mind/brain consciousness to full body awareness. Learn this, feel it living within you and feel it repeatedly.

Mentally oriented people have no concept for living in their bodies. The day after San Pedro, interest in the internet, intellectual conversation, and newspapers, drops to a low level. Alertness is high, fear is low, and thinking is low. This is a huge Gift of San Pedro. Receive it fully!

Therapeutic Effects of San Pedro

My sense now is that San Pedro is a miraculous medicine that has the potential to cure most of the world's woes. Perhaps this is why it is being sold by Home Depot across the United States. What is true? San Pedro is a powerful medicine, no doubt. However, intention and expectation play a major role in the outcome.

The simple distortion of reality provided by a mild hallucinogenic adds to the credibility of the medicine. You have identified one of the principal ingredients, that of strengthening the etheric template, squeezing out distortions, and often bringing them to the light of awareness. Elimination and suspension of the mental processes is another big key. Getting out of the way, to a large degree, is a suggestion that opens people to the voice of God within.

Is there a specific teacher associated with the use of San Pedro? Yes, Me. When illusion is stripped away, I am still there. There is a quality of grounding, of becoming one with nature, which is a big part of the healing. Linking up with, or remembering your true roots is a big part of you, of humanity, and of Earth becoming whole and complete.

Just as by eating the flesh of animals you become them, or absorb their emotional/mental processes, and eating vegetables brings you to a different, more peaceful state of consciousness, taking San Pedro fills you with the qualities which we are discussing.

A very strong etheric template is important to the survival of the San Pedro cactus. It endures very moist, very dry, hot, cold, particularly long periods of no moisture which would kill most plants and animals. The San Pedro cactus goes into a state of suspended animation, during which time is suspended and the energy is put into maintaining the etheric template. Adversity, light, damage to its skin, dryness, and lack of sun, cause this effect of a strong etheric template to intensify.

The personification of the San Pedro, talking about what the cactus "wants" to teach you today, is mostly due to the shaman or the ceremony participant being aware of what is "coming up" or being squeezed from the etheric template. While it is unique for each individual, a juicy issue coming strongly to the surface for one individual will tend to evoke, or bring up, similar issues in others.

How important is applying the Vital 1, Infratonic, Infratonic S, and Scalene Light, in association with the San Pedro ceremony? The Infratonic may be of some value in decreasing discomfort during and after the ceremony. This has value in increasing the comfort of the overall intervention. The Vital 1 provides a small effect of strengthening the etheric template, whereas San Pedro provides a very large effect, thus, the Vital 1 is of little value during the ceremony. On the airplane, traveling to the ceremony, the Vital 1 can be of good value in reducing stress and conserving vital energy.

The Scalene Light greatly enhances the effect of the San Pedro ceremony by dissolving mental structures before, during, and after the ceremony, in the space, the temple, in the etheric field surrounding the participant, and within the body or etheric template of the individual. Thus, we have identified 9 different areas of valuable application:

Cleansing the Body Temple, Aura, Etheric Template

Cleansing the body temple of Etheric Sludge, protects the ceremony from residual and floating thought forms. This provides a context, a crucible, in which condensed Etheric Sludge can expand, can easily come to the light of consciousness, and can dissipate. This is similar to washing clothes. It is important to start with clean water, change the water after the wash cycle, and have multiple rinses, each with clean water. Clearing the space with the Scalene Light accomplishes this.

Clearing the aura is the same as for the body temple, except that there are times to cleanse and times to cook. You don't want to run the Scalene Light continuously because issues bud, grow, and bloom, drawing toxins out from deep in the etheric template. With a clean environment this can take 30 minutes to sometimes 3 weeks. Cleansing the aura with the Scalene Light will generally increase the amount of material released in the ceremony, but in some cases can reduce the "bringing to consciousness" phase.

The etheric template cleansing is a very different application from the body temple and aura cleansing. When someone experiences deep physical and emotional discomfort in a particular location in the body, this is because of an unwillingness to let go of the issue at some level, or more likely, that the sludge is embedded or tangled in the fabric of the etheric template. A deep cleansing with the Scalene Light of the local area of discomfort will act as a lubricant, making the sludge temporarily transparent to the etheric template so that it can move out, unimpeded by the entanglement. This can not only relieve pain and discomfort, but can also allow the participant to "move through" the release of long-standing issues in a few minutes, which might otherwise require several ceremonies. In addition, smaller doses of San Pedro can provide greater clearing, and allow a greater level of conscious participation in the process.

An analogy might be, rather than setting the washing machine to maximum agitation to scrub out stubborn dirt and stains, deep cleansing with the Scalene Light is like adding a good detergent that unlocks the stain from the fabric so that it can more easily float away.

It is important to follow a deep cleanse with the Scalene Light with an aura cleanse from a distance to disperse the Etheric Sludge released into the aura by the deep cleanse. The deep cleanse is best done outdoors when practical to minimize the effect on other ceremony participants. Finally, where issues come to a boil over days or weeks after the ceremony the Scalene Light can again be used for a deep cleanse.

How do we know when the San Pedro boil is fully ripened? The answer is that, when the participant shows up and requests or feels like a deep cleanse would be good, this is a good time. The deep cleanse allows the Etheric Sludge to escape, softens even deeper layers of sludge for release, and generally avoids the outcome in which the body stuffs the uncomfortable sludge deep in the etheric template again.

The Infratonic S is a very different tool, which is used in a very different way. It is generally used to target unconscious limiting beliefs or other fields of dense emotional constriction. It is generally used before the ceremony or between ceremonies, though might be used during a ceremony, where specific clearing objectives are pursued. For instance, someone with a high degree of post-traumatic stress, or someone who specifically seeks liberation from unconscious limiting beliefs, might use it extensively during the ceremony, particularly 10 to 20 hours after taking the San Pedro. San Pedro cactus is like the washing machine – Scalene Light is like the detergent. The Infratonic S might be a prewash stain remover.

Etheric Sludge Warps the Etheric Template

So, how does emotional, mental sludge mess up the physical body? The physical body has an etheric template made up of electric, magnetic, mental, and other substances, which direct the atoms, cells, and organs toward healing and homeostasis. Emotional and mental stuff warps this etheric template, making it direct the body to do strange things, from muscle tensions to inflammation to tumors etc.

They cause it simply to malfunction, holding physical abnormalities in place that would not be there without the worries, wants, and preconceived ideas. "I am separate," "I don't deserve," "this is overwhelming." These are astral mental structures, which play havoc with the etheric template. They are like a poison, causing the etheric template to warp, to start producing substances and structures that don't normally exist in nature.

The Infratonic and Mobile Medic, to a large extent, break up these astral patterns in the etheric template. The Scalene Light works at a mental level, breaking up higher-level structures, influencing health less directly, yet more profoundly. The Infratonic S is effective where astral material has become embedded in the etheric template because when it gets embedded in the etheric template it sets up a perfect vibrational defense for stealth survival in the physical body. We often don't even know that these limiting beliefs are hidden in our etheric template. These are generally structures that have developed over many years or in most cases, millennia, having found ways to embed themselves in the etheric template and defend themselves from vibrational waves that normalize most other astral and mental energies in the etheric template.

The etheric template has specific ways that it remains intact and coherent despite all the chaotic activity in the world. The most difficult structures to remove are the astral and mental, as they weave themselves into this protective matrix, so that the normal etheric winds in nature don't blow them around. The Infratonic S creates a different sort of etheric wind, a coherent wind, if you will, that blows through the etheric template and blows away astral patterns that aren't supposed to be there.

Single frequency electromagnetic disturbances, cell phones, microwaves, 60 Hz power, etc., and fields of emotional and mental congestion, all disrupt or jam the etheric template in a temporary way. Long-term exposure is the problem because this causes the etheric template to begin to lose coherence.

The Vital 1 with a San Pedro Seed

We have discovered that the Vital 1 stabilizes the emotions in a big way while solidifying the etheric template. Put a seed from the San Pedro cactus into the plasma chamber to enhance the effect.

The Vital 1 strengthens the etheric template protecting against outside vibrational influences, and drives out internal influences of emotional and mental nature, which are influences that do not belong to the etheric template. This effect is huge in that stuff embedded in the etheric template is the cause of most chronic disease.

How important is the seed? The seed is two to four times as effective as the Vital 1 with the green crystal. Just one seed is optimal. What about the glue? Superglue will work fine. The crystal was a placeholder. We have been waiting for the day that the holographic projection of the seed would be placed in the plasma chamber to strengthen the field of the wearer.

How long will the seed last? Tell the Deva of San Pedro of your plan. Then the seeds will be good for 100 years. All of the world's San Pedro cacti and the heavenly hosts will maintain the integrity of each seed. *What is the ideal location for the placement of the seed?* Next to the chamber near the magnet. This is an extraordinary technology.

[*Reader: This is important to note in all aspects of our lives. Our intention and prayer call in unseen spiritual beings or beings of consciousness, and they act in our world on our behalf. This is a huge difference between modern science and shamanic science. In modern science we work alone. In shamanic science we work with the consciousness that is present in all things.*]

The Vital 1 is complete. There is no need for three signals. They remain a potential, but will be more of an obstacle to sales and service.

The Vital 1 is extraordinarily effective at projecting the etheric template of a sample such as a seed into the energy body of a person's "electric field." This is where the sample is a living seed, a rapidly growing, or healing seed.

The San Pedro ceremony, with its emphasis on attunement with nature, creates a hyper coherent etheric grid that heals the etheric template, ejecting astral, mental and residual stuff that had been woven into the etheric template. The ceremony holds the etheric template strongly in place and identifies those interfering astral and mental structures, bringing them to consciousness.

The Vital1 has a similar effect, stimulating the etheric template such that it becomes more coherent and tends to reject interfering programs. The Vital 1 also does some internal clearing of this astral, mental material, which is what some users feel and describe as an energetic pulsing of the Vital 1 in the body against areas of congestion. But mostly, it intensifies the etheric template so external fields don't penetrate.

The difference between San Pedro and ayahuasca, is that San Pedro, like the Vital 1, tightens the etheric field, and San Pedro tunes the etheric field into nature. Ayahuasca, instead, ups the DMT and holds it up with

MAIO inhibitors. This opens the door to all sorts of spirit beings and suspends the critical judgment, essentially opening up a space of spiritual consciousness. *[I have not tried ayahuasca, and for now, don't intend to. I am not a hallucinogenics fan. I was just drawn into San Pedro because it entered my life as part of the land I purchased... Or did I purchase the land to meet San Pedro?]*

Feeling Vilcabamba

We have awaited your return. So much to complete. So much to add to your work. Notice the soft flavor of the air, of your skin. All aspects of Etheric Sludge are lower here. With the stronger ozone layer aloft, reduced cars, industry, and fluorescents, and increased abundance of forest, the UV is much, much lower. The radio frequency and electromagnetic emissions are much lower. The sound of human thoughts and emotions is much lower. It is still important for you to cleanse with the Scalene Light. It will be a deeper cleanse since you are starting out in an environment that is so much cleaner. Do it now.

Feel how you feel different now. You are more aware of your back, of your posture, and of your energy. This morning, I would like you to write of how you feel in Vilcabamba.

I don't like the feel of the road as cars drive by. The silence feels so much more delicious. Perhaps I would not enjoy the silence so much if it weren't punctuated with car noise. The noise fades into the sound of the wind. I hear no birds this morning.

I feel a degree of congestion in my 3rd eye. That was your straining for the slight sound of birds. *This is a quieter house than Matt Monarch's dome house.*

I am aware of my tensions in my shoulder and back. I can feel the energy come alive as I relax and sit up straight. I can feel the energy grow and spread in my back as I relax. I feel at peace.

The water feels different. Smoother, friendlier. When I arrived I drank some of the water I had brought from Quito airport, and it tasted chalky, chemically. I hadn't noticed the flavor of the water while at the Quito airport.

Vortex on the Land

Your work with the land was stellar. The land has been waiting for you and your timing could not be better. The Vortex on the land opened wide just in time for the alignment at the Solstice. The energy of the Sun poured through to greatly amplify the work of opening the vortex. The result is a vortex that is 10 times as big and 100 times as potent. The indigenous know what happened and can feel the power. Expect them on the land to bless the abundance over the next 2 to 3 days. They will watch over you and the land and inspire greater works and achievements.

[And I thought I was just buying some land in Ecuador.]

24. Choosing

"Choosing breaks the pattern of dwelling."

4:00 AM

The problem of diet is one of personal choice to take one's life into one's own hands. Each individual must choose. Each individual must hear my call and answer by realizing that the din of illusion created by the capitalist system is nothing but a heartless cloud of self-serving illusion created by so many millions/billions of entrepreneurs, seeking to get ahead.

Hearing my call involves first realizing that very little easily available information is good for the individual. It is really difficult to tune it out, and yet, once it is tuned out and the field is clarified, the value of tuning it out becomes not just obvious, but imperative.

Once this stage is reached the field clarifies further and the vision of Truth versus self-serving "truths" emerges. Every choice made becomes critical. Every time one eats processed foods, factory meats, even veggies grown in congested urban areas, they become "heavier," "denser." Their fields become clouded, and they become easily misled, guided toward servitude to capitalist wanting.

Everything one sees and hears, particularly the easily available media, newspapers, TV/radio news commentary, movies, TV programming, and all sorts of advertising and propaganda, plant seeds of confusion, self-questioning, and a belief in external authority. It is very difficult to maintain a clear field of judgment when any of these information channels are open. It is an individual choice to change channels. Peer pressure seems to push us to watch the latest news, sports, movies, and TV, and to eat those things which cloud the field and cloud the mind. One must be strong to be able to "stand up," against the vast majority - and yet it is a private, personal decision to stand up against the illusions spun by the capitalist system, and to stand in the clear Light of Truth.

Capitalism is not a curse. It brings a clear and simple choice: To stand up or sit in the mud. Past methods and systems brought more difficult choices; dictatorships, tyrannies, slavery, communism. It used to be that standing up brought burning at the stake, ex-communication, or a knife in the back. Now someone who chooses not to watch TV, read the paper, or eat processed foods – is left to their own choices. There is a gentle peer pressure to get individuals back into a clouded, easily influenced state, but it is gentle, and is mostly self-generated. "What will they think? I'm ordering a salad. They think I'm being unsociable because I'm not eating the hamburger. Not going to their church, not buying their products, not repeating or even believing what they say."

In this amazing new world after 2012, everyone has free unrestricted right to stand up and to see clearly. This is a simple choice which immediately brings lots of old, limiting beliefs to the surface where they can all be cleared.

It all starts with a simple choice. "I am fully responsible for everything that happens in my life. What I eat, what I watch and listen to, who I associate with, what drugs and nutritional supplements I choose to take, what level of abuse I am willing to suffer. From this moment forward, I am 100% responsible for every choice I make; conscious choices, unconscious choices, and those choices I choose not to make, in which I choose to allow others to choose for me. I am 100% responsible for all of them."

When an individual stands up and chooses to take full responsibility for every one of their choices, every minute, they are on the pathway to Heaven on Earth. It may seem like an uphill battle, but in fact, not making choices is much more of an uphill battle because you are choosing to be a slave to other people's choices.

The big challenge arises when you are faced with a choice between what you know is true and what your suspicious limiting beliefs tell you. "I don't deserve," "I will be punished," "I don't have permission," "I am separate from God." Navigating these limiting beliefs seems really painful because of another subliminal belief: "It is best not to know, not

324

to ask." You may find yourself cultivating a space of ignorance, somehow believing that it protects you, but it is far more painful not to ask, not to know, than to pursue answers and to make informed, difficult choices.

Another big one is "victim." You have, within you, beliefs that you are a victim of the actions of others, whether it be salesmen, churches, government, family members, or employers. A "victim" is one who has chosen not to make his/her own choices. In truth, there are no victims, no perpetrators. There is only you, making every choice, and creating your life one choice at a time.

Make every choice as consciously as possible and walk through the fire of your own unconscious limiting beliefs. You will find your life rapidly evolving toward a heaven of your own creation. I will be with you every step of the way. Know this is true.

Choosing to Create Uncertainty

You appear to have a great deal of work ahead of you now. In truth you always have an infinite collection of tasks to choose from. Currently the tasks are a bit more organized. All will not get done.

Choose. Choose what is most interesting to you. Everything else can wait or be ignored. Some will get done. Most will go in other directions. This is just as it should be. Enjoy every moment.

Focus on whatever's next. Focus on your choices and choose. If an area of your life has no choices to be made in this present moment, skip it for now and focus on an area with current choices. Make choices and move to another area where you perceive choices. In truth your world is so full of choices that you can make 10 choices a minute for the rest of your life. Conscious choices. Choose nothing out of habit. Choose the unexpected, the unconventional, the unknown. Choose to make a difference, to change the world. Choose love and joy and bliss.

Choose to pursue excitement and to create excitement. So much to choose. You choose the topics of this writing. I provide lots of interesting options to choose from. Always choose what is interesting, exciting to you now. No need to think, to analyze. Choose what feels right and "Move On." No need to "see what happens." That's anticipating and becoming tied to your thoughts of the future.

Choose to choose your new life rather than inheriting your old life by not choosing.

Come from a base of tranquility and love. Make choices gently, and let them go. Move back to the tranquility and again select from the myriad of choices. Always choose love.

Choose the Unexpected

The joy of life is amplified every time you choose the unexpected. Choose to break patterns. So much unnecessary structure exists in your life and in your world. It must crumble. Choose to pull your power from it. Exciting creative opportunities and experiences will emerge as you pull your power from the old. Only when people withdraw from repressive governments, banks, industries, and religions do they perceive new possibilities and begin to make innovative choices. There are hundreds of new innovative choices to choose from every minute.

As you pull power back from your "world," you are freer to live spontaneously on Earth, making actual choices rather than remaking past choices or "playing out" anticipated scenarios. You are not your future dreams or your past achievements. You are here now making choices. Pull your power back from unconscious and repetitive choices. As you pull back your power from your world your field will expand. Everything about your life will expand. Pulling back your power from your world is a huge way to expand into love and joy.

If ever you feel you don't have choices, pull back your power from your prison, your 'self-created' world. Don't discriminate bad or good. Pull back your power from all that is "established" in your life.

Who said that you had to suffer in order to survive? So long you have suffered in the belief that you needed to work; to suffer, in order that you survive. Was it true? It appears that the exact opposite is true. Suffering begets suffering. Joy begets joy. Love begets love. Joy is a much stronger indicator of survival than suffering.

You Were Never You

It has been so long since you've relied upon understanding for decisions. You rely on trusting me and the confidence that whatever you do is perfect. This is a huge shift. In the last few days you have made many big decisions which would have been difficult for you 2 years ago.

You are so different than you were. Your friends are so different too. *Why?* The answer to the puzzle is that the world is so different. Has everybody noticed? Many have chosen to minimize and marginalize this experience. Thus, a survey would find that most people, and certainly the news media, have "decided" that 12/21/12 was a non-event, insisting forcefully that anything like what you have written couldn't be, so it isn't so.

Welcome to my world, a world of possibilities, awaiting realization. You are on the threshold of full immersion, hesitant to jump in. *Why hesitant?* Certainly there is Etheric Sludge. Specifically, you want to know that you will still be you. This is, of course, impossible. *You were never you.* You have always been Me masquerading as what you think is a good, obedient, or proper you, repeating it day after day, and breath after breath.

Continuing to be you is, of course, silly. It's a game. Nobody really cares if you are the same or different each day. Most, in fact, would enjoy it if you were someone different every day. The benefits of assuming different personalities each day are huge. You will be sending off your old habits each day.

You are afraid you will forget how to be you. This is the whole point. Why not be Sekhmet this month? You see the benefits. Why not be Thoth or Leo or other Signs? Why not be Me?

You, in fact, are being Me. Just as, when you drink San Pedro, you become San Pedro, when you start writing in this way you become Me.

You are thinking, "That would be fun, pretending to be God, or Thoth, or Alma." However, you are thinking that you are "really" you "pretending to be" someone else. Practice letting go of Richard every day, a little or a lot. It is liberating. *How can I not be me?* Start doing things in new ways. Start doing new things.

But I like morning writing, eating veggies for breakfast, lighting a candle to Alma, caring for the garden, shopping for veggies, traveling to Vilcabamba, walking on the beach, talking with strangers, calling up friends, playing volleyball, and taking a long, hot shower.

What you do and who you are are very different things. You can be Me or anyone else and continue doing those things. In most cases you will do them better. Morning Prayer, reverence, appreciation. Each thing you do can be so much more reverent and coherent. This is a big opportunity. Adopt a different attribute every day and become it. Do the things you want to be doing every day, but do them as a King, as a devotee, as an empath, as a Leo.

Choose an attribute and exaggerate that attribute. It doesn't matter how others respond. Be you being someone else, and let the chips land as they may. Observe how your world is different every day. Choose people whom you like. Emulate them. Become those qualities that you like. Enjoy being them. Love being them. Feel the bliss of being them.

Change everything until nothing is the same. No need to do anything. Change your schedule on the fly. Be religious about changing your schedule.

Teachers and Intention

Teachers do whatever it takes to get the student to step out of the box. Great teachers don't need to know what they're doing. They only know what is needed for the student, and the teacher is just another student making choices. However I guide the actions of the teacher and the perceptions of the student such that the desired illusion (or impression) is created within the student.

There is no right action or choice. There is no right thought, feeling, or belief. It is all about broadening your perspectives, releasing the limiting thoughts, beliefs, fears, and wants. Know that there is only you having an experience with you; playing another role.

Every aspect of your life is a lesson. The opportunities for expansion are endless. Each individual does it differently. Each event, no matter how similar to another, is unique. So many subtle variables are different. Don't worry about which you do, what to feel or think. Just choose in the moment and let it go. Life on Earth is easy in that way. So much work and joy in the life ahead of you. Choose and Move-On!

It is choice that resonates throughout the Universe. It is the choice itself that is the seed of growth and expansion, not the circumstances surrounding the choice. You are making choices and, for the most part, moving on. This is setting up a huge pattern of expansion, breaking one million strong patterns of dwelling on a particular situation and not making a choice. When you "do well," you are dwelling in an illusion. When you choose, you are breaking the patterns of being stuck in dwelling. Fleeting from dwelling is a big step forward, particularly when you observe the dwelling and how choosing breaks the pattern of dwelling. Often the most powerful choice is the solid forceful choice of no action, then flitting to another flower.

Follow your attention and you will find a myriad of flowers, some more fragrant than others, none which serve you to give more than a fleeting dwelling. The most valuable choices have been those which keep your life moving, which break stuck patterns, and which lay the foundation for more informed choices in the future.

Flitting among the flowers! All that attracts you is a flower with its beauty and its nectar. You wouldn't appreciate the fragrant nectar so much if you weren't flitting among sour, bitter, bland, too sweet, and a million other varieties and textures. It is not about choosing a favorite and making it your home. You are your home. Your home is the entire universe and more.

It is all about observing your choices...........

and moving on to the next flower!

Stepping Forward to Choose

You are welcome to wallow in the past if you so choose, but we ask that you choose it consciously, to be aware every minute, with every choice, that you are choosing to repeat a past experience.

Laura Doubts her Abilities

Laura: You know, Richard, sometimes I get so sick of it all. Like I woke up this morning, observed my dreams, and realized the things my dreams were revealing to me, which was all pretty accurate. I felt bad for sabotaging myself and my life all these years due to insecurities, which resulted in loneliness, depression, heartache, dissatisfaction with work and being broke. I feel like I'm working on it now and it's probably getting better. But is it really? And isn't it sad that someone at this age "might be" just figuring it out now? Especially when I see people 20 years younger than me and many of them have figured out already what's important to them and what they want. I really envy them and wonder why I have been the way I have. Although when I was 20 I thought I had it figured out but didn't have the guts to go after my desires or I didn't think I was allowed. Anyway, I'm just babbling on expressing my sorry feelings for myself with a friend who I think is kind enough to care and perhaps give me some wise advice. I know you might say pull your power from it. But how many times? And does it really work? Sorry to be sounding a bit pessimistic this morning but it's one of those days.

Richard: You wrote: *"I woke up this morning, observed my dreams, and realized the things my dreams were revealing to me, which was all pretty accurate. I felt bad for sabotaging myself and my life all these years due to insecurities, which resulted in loneliness, depression, heartache, dissatisfaction with work and being broke."* That is spectacular insight into how you have created your life up to now, and an opportunity for you to choose something different every minute.

You wrote: *"Isn't it sad that someone at this age "might be" just figuring it out now?"* Is there any time other than now? Is there any better time to figure it out than now? How can you really, in good conscience, claim that it is sad to be figuring it out now!!!!!!

You wrote: *"When I was 20 I thought I had it figured out but didn't have the guts to go after my desires or I didn't think I was allowed."* Do you know what your desires are? Do you now have the guts to go after them? Do you now feel that you are allowed to go after your desires? Who is allowing you? Who is not allowing you?

Is it possible that what you really want, what you really desire, is to wake up and discover that this is *YOUR* life and that you can make of it what you want? This is not your parents' life, or the life of your church, or the life of your ancestors, or the life of your family, or the life of your culture. What you have wanted all your life is to wake up and realize that this life is yours and you can make of it anything you want. Maybe one day next week you will wake up and realize this is true. It's like sitting in the back seat of your car all your life, then one day, realizing that there is a driver's seat, then sitting in the driver's seat and starting to drive your own car. You are welcome to think of me as sitting in the passenger seat observing you huddled in the back seat complaining that the car is not going where you want to go.

You wrote: *"I'm just babbling on expressing my sorry feelings for myself."* You are observing yourself in the process of creating a world in which you can feel sorry feelings for yourself. You are an extraordinary being, capable of creating remarkably believable feelings of misery. You are Amazing! You are an amazingly powerful creator! You even really believe that you are miserable! Amazing!

You wrote: *I know you might say "pull your power from it." But how many times and does it really work?* Jesus said, "Go forth and sin no more." Perhaps he meant to pull back your power from those remarkable feelings of misery you have created, and don't create them again. It's one thing to pull back your power from a creation that you don't really want, and another thing to use your unconscious creative power to put those things you don't want back in place. You are truly amazing, particularly with your ability and tendency to pray a lot.

One morning you will wake up totally miserable, and then suddenly come to the realization that you really can pull it all back, that you really can choose to create the life that you really want. Then, in one breath, you will pull your power back from all that misery you felt that morning, and know that it is true.

Trust your higher self to continue to give you insightful dreams such that you wake up in misery and with the conscious awareness that you are able to pull back your power.

Conscious Choosing is Descending

How long do we continue like this with me writing in the early morning? It is certainly your choice. My suggestion is that your best choice is "forever." There is no limit to how far you can go once you are in communication with the All.

Remember to maintain confidence and humility. Know that, from the standpoint of your earthly realm, you are called to command. At the same time, you are nothing but a shell, a vessel for Me to occupy. It is a part of the vessel that feels itself separate from the All. This vessel can easily come to believe it "deserves" everything it can demand or take. This is the delusion of the ego. Remember that you are a King because you are Me and it is your duty to be a steward of all life on Earth.

It is your nature to experience love, joy, and bliss. It is not your right to demand these things, but rather, what remains after you get the Etheric Sludge out of the way. There is no need to command or demand love, joy, or bliss. It is delusion to do so. Your nature, once you clear your limiting beliefs, is to have all your needs met by Nature. Don't assume that I will take care of you. Yours is a world of choice. You create your circumstances every minute. If ever you get to the point where you expect me to rescue you, you have created the delusion of being a victim, of needing to be rescued.

It is all of your choosing. There is nothing more real in your world than your choosing. Even what you choose is less real than your act of choosing. In fact, the you that you consider yourself to be is incapable of choosing. You can only react, repeat the past based on recollection of old ways, or do as you were taught. Whenever you step up and make a truly conscious choice, it is I, within you, who make the choice. Thus conscious choosing is a means to Descend, to become Spirit walking the Earth.

It is not simply a matter of "being in body." It is an endless progression of descent and expansion. Every time you choose consciously you become more real and what you choose becomes less real, and simply the outcome, the product of your choosing. You can choose to create some aspect of your life for years or decades, then, with one choice, walk away – withdraw your attention and support, and that part of your world vanishes as if it were never there. You can, of course, choose to re-create, to bring an old creation back into your world. But why? It is true that it seems easier the 2nd time or the 10th time, but the fact that it seems easy is just the fact that you are choosing an unconscious pathway, making past choices to remain in the illusion of the past.

You will find joy, love, and bliss each time you choose a new world of your preference. Every minute your life is refreshed because you choose it now. You are welcome to wallow in the past if you so choose, but we ask that you choose it consciously, to be aware every minute, with every choice, that you are choosing to repeat a past experience. From this perspective, you can see what a folly it is to try to please someone else. All you find yourself doing is trying to repeat past choices that seemed to please this person before. This is living in the past. What others love about you is not your endless desire to reach into the past and to choose old behaviors that "seemed to cause pleasure" last time. Rather, they love freshness, renewal, innovation, unexpected surprises, *ANTICIPATION* of the unknown.

This anticipation, this uncertainty, is why you came to Earth in the first place. Trying to repeat the past is the disease. In fact, it is the cause of all disease. Yes, all disease can be eradicated every minute, in the now, by choosing it so. It frequently takes a little time to eliminate the metabolic byproducts produced by the spontaneous remission from disease. Just as in the mouse that recovered from cancer, the remission stems from a single choice then proceeds to materialize over time.

Choose anew, afresh every day and you will leave yesterday behind and step into a tomorrow of your choosing, filled with joy, love, and bliss, because it is filled with your presence, which is, by nature, your right to live in Heaven on Earth.

25. The Luminance of Kirlian Photography

This chapter provides insight into the mechanisms behind the movement of vitality in the body by explaining what causes the brightness of the Kirlian image and a mechanism by which energy is transported in the body. It explains how a belief system like "I am tired" can cause our vitality to vanish. It addresses why Qigong Masters often have no Kirlian image, and what causes the brightness of the Kirlian image to vary with the vitality of the body.

I have done much research into this puzzle over the past 25 years, taking tens of thousands of Kirlian photographs of people in such circumstances as during long, intercontinental airplane flights, and in cancer hospitals in China. I found that cancer patients generally had no Kirlian image, and when I applied the Infratonic to their feet for a few minutes, their Kirlian image, while brighter, only showed streamers, long thin lines like lightning created as electrons, stream to the fingers along ionized pathways. Here is the basic Kirlian layout:

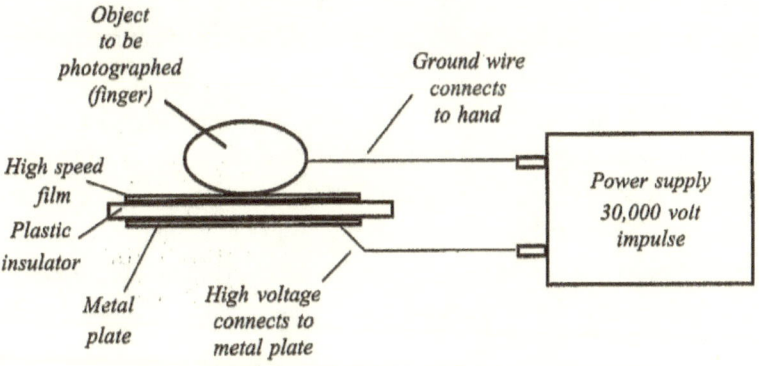

(The Kirlian camera applies high-voltage behind the film that the fingers are pressed against. This high-voltage causes what is called a corona discharge, a discharge of electrons which illuminates the air as the electrons strike and excite the air molecules.) These Kirlian images showed no balls, which are otherwise created when as electrons move away from the finger tip. This was a big puzzle to me.

In addition, I found that, after a few hours in the cancer hospital, doctors and visitors also showed reduced brightness of the Kirlian image and only showed streamers. It was as if the cancer itself was sucking vitality out of the patient, doctors and guests.

During this time, back home, I noticed groceries appearing in front of a neighbor's door. I knocked and discovered that she had undergone cancer surgery two weeks earlier, and was not recovering her energy. I immediately offered to take her Kirlian photo, which started a week long research project. The first day, she had virtually no Kirlian image and I gave her an early prototype of the Vital 1 to keep with her full time for a week. Each day we took another Kirlian photo and each day her photo got brighter until, by day 6, her Kirlian image was normal. She told me she had gone back to work that day, and felt fine.

From all this research I knew that the Kirlian image was a good indicator of the vitality of the test subject. Many researchers over the years have studied Kirlian photography and have come to the same conclusion. The problem with Kirlian photography, and why it is rejected by mainstream science, is not because it is inherently full of uncontrollable variables, but rather, because it succeeds at measuring the vitality of the body, and because the vitality of the body moves as a result of intention (an uncontrollable variable). I have tested a few people who are consciously able to make the Kirlian image of their hands brighter or less bright depending upon their intention. Most don't have this conscious ability, but all do have the ability at an unconscious level to move vitality. It is this factor which makes Kirlian photography hopelessly uncontrollable to most rigorous scientists.

I spent many years adding controls to my Kirlian cameras to reduce variability and increase reliability and consistency of results. The ground electrode is probably the most valuable addition. It eliminates capacitive conductance effects in which a test subject will get much brighter images when her foot is moved close to a metal object, such as a filing cabinet or a desk, and gets dimmer images as her foot is moved further away.

As I eliminated uncontrolled variables, I was thinking that I would find that the electrical conductivity of the fingers changed according to how much light was produced. I expected that, when the streamers but not the balls were produced, we would see a higher electron flow into the finger than out of it. Surprisingly that's not what I found. As I measured the current flowing in and out of the fingers, I found that it did not vary. People would have no Kirlian image at all, would then go outside into nature and breathe with the intention of collecting Qi for 10 minutes, then come back and be tested again. To my surprise, the electrical currents in the fingers were exactly the same, but the light produced by those electrons flowing through the air, as recorded on photographic film, was much brighter after they did their breathing exercises. What changed? Why were the same electrons producing more light after these people intended to collect Qi?

This puzzle had been haunting me for many years. The following is what I wrote through "automatic writing" at 4:00 AM one morning a few weeks ago in answer to these questions. If it proves true, it will be revolutionary for modern medicine and human performance.

4:00 AM

What is the nature of the change in brightness of the Kirlian image from such actions as massaging another person and deep breathing outside in nature, with intent to collect energy? Just as the bottoms of the feet can be opened to release congesting energies into the earth, making the Kirlian image bigger and brighter, intention to collect energy causes the same.

What is the nature of this collected energy which makes the corona discharge more illuminated? Vitality, that form of energy which animates and motivates, also illuminates when electrified. There is no limit to the availability of this illuminating vitality in a space. However, vitality follows thought so that expecting to "lose energy" usually makes it so.

Air Molecules can Collect Vitality for the Body

[Cancer patients would tend to get less and less illuminated Kirlian pictures as the cancer progressed until they had no pictures at all. The pictures that they did have would show streamers and not balls. This suggests that the electrons from the environment coming into the fingers have this elusive illuminating quality of vitality, and the electrons coming out of the body do not have this vitality. Further, as the cancer gets stronger and the patient gets weaker the vitality is even absorbed from the air or field around the fingers, so that no light is emitted during either the flow of electrons toward the finger or the flow of electrons away from the finger. This finding opens a whole new area of questioning.]

Why is it that the cancer has abundant vitality and the rest of the body, and the area around the body is depleted of vitality? In cancer, the cancer views itself as a dynamic source of energy and the rest of the body views itself as "having cancer." This is a perspective that causes the body to collect energy and give it to the cancer. All concentration and scarcity of vitality are brought about by intention.

What is the nature of this vitality such that it produces measurable light when stimulated electrically? There is an interesting relationship between vitality and electrons. Electrons have intention, so they can transport this vitality.

Are they carriers of vitality? No, not as you view it. They may fly through the air crashing into air molecules causing the electrons to be lifted to higher orbitals then to fall back, emitting light. Alternatively, they may come from a packed, deficit state within an atom into outer orbitals, and then transmit that energy telepathically to another place where they believe it is needed. Thus, instead of emitting light through spontaneous falling from the outer orbitals, the electrons move to lower orbitals in a way that transports usable work, vitality.

This is an often-ignored mechanism because it is related to intention and because it is not generally measurable except in Kirlian photography.

338

The ability of an atom to transmit this energy of a descending electron is a huge mechanism, overlooked by your current science. It is transmitted as a phonon, a coherent unit of etheric energy imparted to a coherent wave function. This is all mediated by intention. The atoms in air molecules can also store this energy by choosing to hold electrons stable in outer orbitals. These are the easily illuminated atoms seen in Kirlian photography.

The body field can call for air molecules to be collectors of this energy, in which case these air molecules dissipate no light. On the other hand, where the body carries an excess charge such as an inflammation, it calls on the air to dissipate this excess energy. It appears that when the body calls on the air around the body to be vitalized, bright Kirlian images result. And when the body calls for low energy in the air around the body, little energy is stored in the outer electrons so little light is produced.

Streamers (photos where you only see light in the electrons moving toward the body) are produced as electrons are galvanized or charged by the electrical pulse, and then release it closer to the body. Balls indicate highly charged electrons moving further away from the body to dissipate excess energy.

[I don't know about you, but I find concepts presented in the above paragraphs to be quite challenging. As I read these paragraphs again and again while editing this book I began to understand more and more of what my engineering-educated brain tends to reject. This helps me to understand part of why it's so important that I write at 4:00 AM when I am groggy, and when my engineering brain is still asleep and not interfering with what is written.]

Dissolving the Disease Signal

[Here's an article that, if it proves true, or partially true, may reduce transmission of disease.]

Today, we would like to speak of the challenge of the ages, cancer. It truly does live in the earth. It is important to clean it up in bodies before burial, generally with burning, cremating the body. In one generation, 20 years, this will eliminate cancer of all sorts. Do you see? Cancer continues to live in the earth as it continues to consume the bodies of its victims. This process can take months or years.

It is now known that the hundredth monkey effect, where monkeys on an island spontaneously learn a behavior practiced by many monkeys on another island, is a real phenomenon. Another proven effect, entanglement, is proven in studies which show that two people with close relationship tend to show synchronized body rhythms. Cancer is a similar effect. As buried cadavers are consumed by expanding cancer colonies, these hungry cancer cells produce a resonant signal that is carried by what is similar to the hundredth monkey syndrome or entanglement. While there are many factors which increase the likelihood of cancer, like carcinogens, the start of cancer involves normal cells which spontaneously start developing cancer in resonance with this signal. Eliminate the signal and the likelihood of cancer decreases.

In fact, lots of diseases are transmitted through resonance or entanglement. Applying the Scalene Light in hospitals will dramatically reduce hospital disease transmission, and contribute to the recovery of countless cancer patients. Just by keeping the rooms and hallways clear of Etheric Sludge, patients will get healthy significantly faster. Applying the Scalene Light around cancer patients in homes, nursing homes, and hospices will contribute to patient recovery, improve quality of life, and reduce cancer incidence in loved ones as well.

Applying the Scalene Light in cemeteries will substantially reduce the strength of the cancer signal. One Scalene Light, aimed upward, operating 3 minutes every 3 hours, will clear up a gravesite of 1/16 square-mile, 200 acres. Within a week of deployment, the environment will be clear of the cancer signal.

The best way to quantify this is through blinded studies of energetically sensitive people, those who feel uncomfortable around cemeteries. First clear the people with the Scalene Light so they are

340

highly sensitive to variations in the energetic feel of the environment and not influenced by Etheric Sludge they carry with them. Then have them travel to several cemeteries and evaluate how much discomfort they feel. You will find that untreated cemeteries will make them feel uncomfortable, but cemeteries treated for a week or two will feel much more comfortable. Many will be able to feel the general location where the Scalene Light was placed as the cleanest, particularly in the first day or two.

In addition to the benefits of clearing the atmosphere around cemeteries, the actual count of living cancer cells in the ground will decline rapidly as well.

[I have no idea if this is true. However, I figure there is a chance. Since this would be of huge benefit to mankind, I have built prototypes for a cemetery clearing device, and am preparing to begin testing as proposed above. Hopefully it will work as described. I hope to put the results in a later edition of this book.]

Breathing in Abundance

[There is apparently more to the science of the luminance of the corona discharge in Kirlian photography. Here is the continuation of the writing started above on luminance, which led me to another biofield measuring device, also under development.]

There is another factor: The push and pull of electrons. A negative pulse applied to the fingertip causes temporary crowding of electrons into outer orbitals of air molecules. The positive energy charge at the skin causes local electrons to move toward the body leaving a positive charge in those electrons of air molecules near the skin. This changes how densely packed the orbitals are near positively charged skin. Lightly packed orbitals produce lower frequency light. Hence, the variation in colors observed in color Kirlian images. As you can see, from this effect there are a lot of variables here. If the atoms are transporting energy into the body, light is only produced when they move away from the body and we see red light.

There is a question here as to whether the positive charges or negative charges are carriers of energy. It is a question of intention and direction. Here we approach the biophoton phenomenon. Where the intention is that the body be vitalized, the observation is that biophoton emission from the skin is reduced as energy is transported to the body. On the other hand, when the skin is wounded, this ability to collect energy is compromised and the energy "leaks out" increasing the bio photon emission.

An advanced Qigong practitioner carries a strong training to collect vitality constantly. Wearing black is symbolic of this constant absorption of energy. This is the opposite of radiance in which the body acts as a source, a radiator of this energy.

At one level, one can radiate and be depleted, or absorb to experience high vitality. This occurs in a belief framework of limitation. "I must take from my environment to be vitalized, or else I must sacrifice myself, my own energy, to help another."

A belief system of infinite abundance allows a person to intend to be a source of infinite abundance. In essence, vitality is radiated into the space around the person. The vitality comes from the void, the quantum space. There is no limit to the amount of energy that can emanate from the heart of each and every atom. Thus, we again return to the belief systems. "Am I a source of abundant vitality? Or am I a consumer of limited resources?" This "abundance" belief system is the key to those people who truly retain their vitality while providing massage, nursing, or any form of care-giving.

"How can I give endlessly to someone with cancer or someone who believes in limitation, who is an infinite sink of energy?" The fact is that a cancer patient is not an infinite sink, but rather an imbalance of high and low charges of vitality created by intention at the cellular level. The whole dynamic of cancer is based on scarcity and allows people to play out all sorts of scenarios of scarcity. It is an expression of scarcity. A shift in perspective toward abundance, oneness, and connectedness, allows the cancer dynamic to dissipate, should dissipation be chosen.

How does one learn to have abundant vitality? Know that this is your natural state. Fear, limitation, and separation are artificial belief systems, which have no place in this world anymore. To have abundance in any form, one must release a mindset of the corresponding limitation. It seems easier to add a new belief or intention than to release an old one, but in fact, nothing could be further from the truth. Relaxing is releasing. The out breath is releasing. Let it go. Breathe it out, and simply don't breathe it back in. Live without the thought of limitation or separation.

You will find that you naturally become a radiant source of abundance and vitality because this is your nature before the belief in limitation and separation was imposed upon you. Know that, on the levels of energy and vitality, you are held. You are loved. You can relax and simply allow limitation and states of low vitality to fall away.

It seems hard, but in fact nothing could be easier. Simply lay back, relax, and trust that you are held. In an instant, in a single breath, you can experience spontaneous remission from all aspects of limitation and suffering.

Thus, the brightness and darkness of the Kirlian image are not scientific phenomena in that they are entirely mediated by intention. Intention stands outside of the art of science as it is currently practiced. Know that you are held, that you are loved, now, always, and forever. Relax into that knowing and breathe abundance.

26. Pulling Back Your Power II: The Etheric Template

"All of your beliefs and expectations are just manifestations of the prison walls that keep you from shining brighter."

Cleansing the Etheric Template

It is time to take a deeper look at the nature of limiting beliefs, to understand them from the standpoint of the structure of our physical body much as residual traumatic stress was stored in the horses' hocks in the equine experiments. The limiting beliefs themselves get stored in the physical tissue, or the etheric template, the energetic body which holds the physical tissue in place.

Some issues causing pain, discomfort and unexplainable behavior are easily removed by the method of pulling back your power. However other issues, those held in place by deeply embedded limiting beliefs, are tougher to dislodge. These limiting beliefs can be from early training or traumatic memories from childhood. They can also be karmic, or those limiting beliefs we bring in from other lifetimes. To experience past life trauma in the current body, we need to find a body and a childhood environment that will prepare us to re-experience these limiting beliefs. Another source of deeply buried stubborn limiting beliefs is ancestral programming. Whatever our great-great grandparents did, our great grandparents probably also did, and so did our parents. Our genetic structure, or the energetic patterns of ancestral limiting beliefs, our ancestral life-stream, is constantly pushing us to do these same things.

Ancestral Guardian

I was sitting in the chapel at the Self-Realization Fellowship in Encinitas on a three-day silent retreat. Each time I sat there and sought to open my heart and blend with the wonderful energy in the chapel, the muscles in my upper back and neck would tighten up and become very painful. I tried a variety of techniques to pull the stagnant energy out of my neck, whether massage, stretching, or shoulder rolls. I closed my eyes and went in there to look at what was going on and found a little man. "Who are you?" I asked. "I am your Ancestral Guardian," he answered. I asked if he could help remove the painful stagnation in neck. He answered that it was supposed to be there; that it was important for the success and safety of my descendants that this congestion exists between my heart and my brain.

Upon further questioning he continued that the reason my ancestors had been successful was because they weren't softhearted fools who would give their hard earned money away. They would set aside their heart, their compassion for others in favor of providing for their family and their descendants. It seemed he had quite a logical and legitimate point. However, I told him firmly that I wanted to connect my heart to my brain, and I wanted his help. He agreed. However, as I continued to attempt to pull out the stagnation, I discovered that he was putting it back in again. At this point I asked him to leave. I told him that his services were not wanted anymore. He was quite tenacious, and even now sneaks in from time to time. This Ancestral Guardian has been a major barrier to my getting in touch with my heart. To the point of this discussion, my etheric template had my Ancestral Guardian deeply woven into it.

This deeply embedded programming is a huge part of our subliminal limiting beliefs. It is hard to remove because it is deeply embedded in our etheric template, the energetic pattern that holds all our cells in place and orchestrates the remarkable healing that occurs each time we get injured. Because ancestral programming is an integral part of this etheric template, or rather deeply embedded such that it feels like an integral part, it is very hard for us to remove it using such techniques as the pulling back your power technique. Our etheric template does not believe it is safe to remove itself.

In this chapter we will first look at the relationship of the sympathetic nervous system to pain, dysfunctional behavior, and the placebo effect. For this, you also need to know a little background on the topic that arises early in the writing. As mentioned earlier, Wilhelm Reich, a psychoanalyst and human vitality researcher from the 1930's 40's and 50's, is the only person in the history of the United States to have his books burned by the United States under persecution by the FDA. He died in jail under false charges. He is referred to in my writing because I work closely with his work with the Cloudbuster, which he used to induce rainfall and direct atmospheric winds. This technology has strongly influenced the development of the Scalene Light and the Infratonic S.

Next we will look at the work that led up to and resulted in, the Infratonic S, a product that came partly out of my experiences with the San Pedro cactus, and methods that might be employed for specific parts of the body to shake embedded limiting beliefs free from the etheric template so that we can more effectively pull our power back from them and be liberated of them. We will finish with an advanced process of cleansing the etheric template.

Sympathetic Nervous System

4:00 AM

How am I doing on life extension? There are still some major limiting beliefs so you're not out of the woods yet. The Infratonic S is very helpful. It is better to keep it moving. One minute per point. Leaving it on one point for 4 hours while you sleep, as you did last night, tends to drill holes, which can be counterproductive. *How so?* It stresses the system and the etheric template tightens, protecting itself and the sludge it hides. This makes it more difficult to get the sludge out in the future. Something like ten minutes per day before sleep, one minute on each of ten different points for a couple of weeks is, in general, optimal for this clearing-of-etheric-sludge application.

What else to write about this morning? Keep writing and it will emerge. Continual flow primes the pump.

It is true that you have, for many decades, been struggling with profound limiting beliefs. This is just how it should be. You are born into Earth with the prevalent limiting beliefs of Earth. Each newly born person comes with a full set of prevailing limiting beliefs, and has the opportunity to release them, or to collect more during a lifetime. Most people collect more and more, becoming rigidified in their unconscious beliefs until they are very stiff and inflexible, physically, emotionally, and mentally. You will notice that you are becoming far more flexible with each passing week. These full moon clearings will greatly accelerate your progress.

How are limiting beliefs related to the persecution of Wilhelm Reich? Reich, from birth, expected to be persecuted. It is a pattern that followed him throughout his life. Or rather, he carried it with him. His reality was that he would be persecuted, and so it was.

This is a part of the theme of his life – to push against established beliefs and have them push back. This is similar to over-using the Infratonic S and creating resistance. Reich's therapeutic methods are powerful Etheric Sludge cleansing methods. Where people have a lot of sludge, and keep hearing about his work, they will often do whatever it takes to stop hearing his theories. Know that his methods are proven and reliable, providing steady progress for his patients and students. However, for highly armored individuals, Reich's writings were, and are like sandpaper on their skin and brain.

Your approach is much gentler. First, you are not telling people that they are screwed up. You're telling them that they are Joy, Love, and Bliss on the inside with a layer, or 10 layers of Etheric Sludge clouding their experience of these pleasurable states. This is a soft approach, an enticing approach. You are providing many methods to reduce and reverse the sludge accretion process. You are easy to follow and easy to ignore.

Why is Reich still being bashed more than 50 years after his death?
First, there are many fanatic supporters who push his teachings, far more
today than 50 years ago. Anyone who pushes Reich's view is similar to
someone who pushes vegetarianism or pro-life or pro-choice. They are
pushing because of Etheric Sludge within them. Once they release the
sludge, they will no longer need to push agendas upon resistant people.
Those who have this need to push against resistance are drawn to
proselytize about their rigidified vegetarian diet or Reich's teachings
precisely because of the resistance they will encounter. Reich's
teachings are polemic. If you want to encounter resistance in life, push
his most friction-producing theories among circles of resistant people.

On the other hand, individuals who want to clear their own Etheric
Sludge will find Reich's teachings to be very useful to accelerate the
process. Explore a new, conceptually challenging field and be aware of
how you resist and oppose it, and you will uncover your sludge.

How is this related to the sympathetic nervous system? The
sympathetic nervous system allows for illusions, belief systems that
contradict known facts. This is most interesting. *Why would such a
system be an important part of the human body?* The answer is most
interesting: Through the millennia, life evolved with direct sensing and
immediate contact with all that is. Plants and crystals are simply one
with truth. Through the parasympathetic nervous system, animals are
fully in touch with this truth. However, as animals evolved, there came a
development of competition among species for survival. Plants simply
know whether they are wanted or needed. They withdraw when they are
not in harmony in a particular location, and grow somewhere else.

Sympathetic System

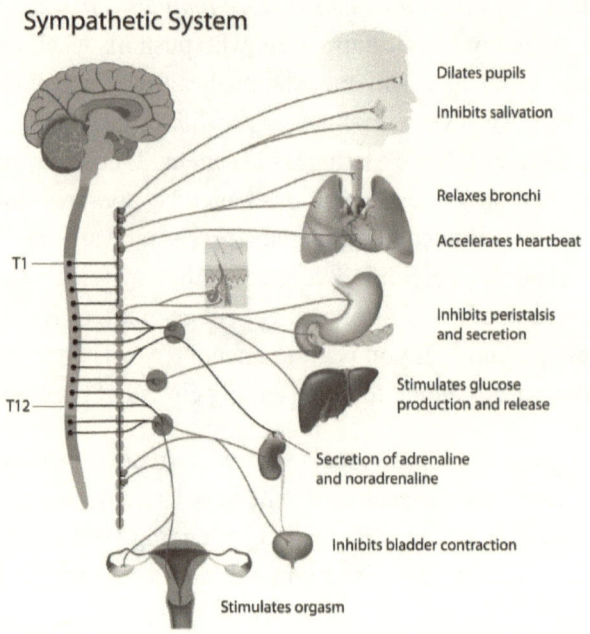

Dilates pupils

Inhibits salivation

Relaxes bronchi

Accelerates heartbeat

T1

Inhibits peristalsis and secretion

Stimulates glucose production and release

T12

Secretion of adrenaline and noradrenaline

Inhibits bladder contraction

Stimulates orgasm

Animals are different. The sympathetic nervous system embodies a fight or flight response. Essentially there are 2 realities; the first being the central nervous system in which there is a universal harmonious truth which all things can follow to feel harmonious. Then there is a personal reality, that of the sympathetic nervous system, in which my well-being depends upon belief systems that are embedded in me. Limiting beliefs are false beliefs that govern our behavior for personal benefit without regard for the harmony of nature. The sympathetic nervous system, driven by these limiting beliefs, causes us to react, often violently, and always in our own self-interest.

It is not that the sympathetic nervous system embodies or contains our limiting beliefs. Rather, it reinforces self-preservation at the expense of the environment. Whatever we perceive to be a danger to our survival triggers the sympathetic nervous system and we are compelled to do something immediately. Destroy the enemy. Become the enemy. Run away. The sympathetic nervous system can be a huge source of conflict as is seen with the teachings of Reich. On the other hand, it can be a rich fertile ground to discover those limiting beliefs, which lead us into conflict and other dysfunctional behavior.

Sympathetic Nervous System	Central Nervous System
Separateness	*Direct Connection with the Creator*
Thinking but not knowing	*Knowing*
Busy brain	*Clarity of vision*
Weak boundaries	*Firm boundaries*
Weak uncertain voice	*Clarity of creative voice*
Grief	*Inspiration*
Fear	*Love and Joy*
Limitation	*Abundance*
Anger, resentment	*Freedom*
Sympathy, Betrayal	*Connection, Oneness*
Fear	*Vitality*
Ungrounded	*Direct connection with Nature*

Reactions of the sympathetic nervous system can most easily be felt in the solar plexus, heart, and throat, those areas of the body along which the sympathetic nervous system runs. By being aware of sympathetic nervous system reactions and exploring the limiting beliefs, which cause the sense of fear or danger, it is easy to identify limiting beliefs and to pull our power back from them.

Here's the relationship between the Power and Presence of God and pulling back your power from limiting beliefs: When your sympathetic nervous system reacts, it tends to control you based on unconscious limiting beliefs. When it doesn't react, your attention moves to the parasympathetic nervous system, the central nervous system that integrates the whole body around the central spinal cord.

Now, and this is interesting, when you react with the sympathetic nervous system you bring about a whole chain of events which lead to unknown results. There are two very different tracks of the sympathetic nervous system, the left side with the spleen, pancreas, heart, and left brain, pushing you to move into your logical analytical mode and to make peace, find a compromise, absorb and be absorbed. This is where the sympathetic nervous system gets the name "sympathetic," for you are driven by the left side toward adopting the belief systems and the limiting beliefs of others for survival. People in committed relationships frequently sacrifice themselves and become like the other person to "save the relationship."

The right track of the sympathetic nervous system is anything but sympathetic. The liver and gallbladder seek to destroy whatever is the perceived source of the sympathetic nervous system reaction, whether a barking dog or the teachings of Wilhelm Reich. The lungs seek inspiration, as opposed to the heart, which seeks harmony and peace. Also, the right brain goes to work providing new alternative solutions which are often game changers and, like Reich, threaten the status quo.

Thus, when the sympathetic nervous system fires, you are pushed to choose between a sympathetic assimilative action to preserve the peace, and a creative out-of-the-box new solution that is likely to upset the apple cart and make people angry as it cuts through the prevailing limiting beliefs of the group. Wilhelm Reich provided methods to engage the central nervous system and to bring people to see the limiting beliefs that drive the reactions of the sympathetic nervous system. This objective perspective allows people to see and unwind their limiting beliefs. However it causes huge fear and sympathetic nervous system reactions in those who depend upon subliminal beliefs in others for their own survival. "If members of my community discover they don't need to be afraid, I will lose control." This is the principal limiting belief evoked so powerfully by Wilhelm Reich. This is why, to this day, powerful people in groups seek to defame Reich. His teachings liberate individuals from their own subliminal limiting beliefs and allow them to think for themselves.

So when you say, *"I am the Power and Presence of God and I pull back my power from this limiting belief"*, from the perspective of the nervous system, you are abstracting your awareness from the sympathetic to the central nervous system. You are also intending to disable, or pull back your power from whatever limiting beliefs are creating the fear reaction that is calling for a fight or flight response.

You are training yourself to choose to remain in central nervous system consciousness. There are many other levels upon which "pulling back your power" functions, but these are a couple of huge ones.

Yes, it is true that there is also a higher consciousness that orchestrates synchronistic events in your life. However, in most cases, the dominant effect, the elephant in the room (the same elephant in the room as creates the pain that can be relieved by the placebo effect), is the hyper-reactivity of the sympathetic nervous system which causes us to be blind to what is in front of us, and the limiting beliefs which cause us to falsely interpret events around us. *[Dr. Zarno shows that the body can induce localized reduced blood flow to create the experience of chronic pain. This is an example of the sympathetic nervous system in action.]* This can leave us completely unaware of easy win-win solutions that synchronicity has put right in front of our faces.

R.:   Thus, pulling back power from the sympathetic nervous system to the central nervous system is a principal mechanism of this process. This is why the Vital 1 with the 2 Hz stimulation is so important. It moves us to a "state of mind" (CNS dominance) in which the sympathetic nervous system is not clouding our vision with reactive fears.

The sympathetic nervous system is not just felt along the front line of the chest and abdomen. It can create dysfunctional pain and discomfort throughout the body. This includes arms and legs, joints, and bones. One popular location is the limbic core of the brain. When the sympathetic nervous system triggers the limbic core, it begins to swell, producing pressure in the brain. Chronic pressure creates tension headaches, and acute pressure causes migraines. Another popular location for pain and tension is right behind the sympathetic nerves, along both sides of the spine on the back, particularly along the cervical, thoracic, and lumbar spine. Emotional issues you don't want to see often show up as chronic tension or pain in the back caused by the sympathetic nervous system. This is why the Vital 1 will seem to "inexplicably" relieve pains. Limiting beliefs and resulting fears through the sympathetic nervous system create nonspecific pain and chronic illness. This lays the foundation for the "placebo effect." Simply by moving conscious awareness from the sympathetic nervous system to the Central nervous system, most pain and suffering in the human body can be relieved. However for lasting relief the power must also be pulled from the subliminal limiting beliefs stored in the local tissue.

In 1920 Paramahansa Yogananda gave a talk in Boston entitled "The Science of Religion" (now available as a paperback book). In it he describes what seems like a very similar system, as a scientific method to accelerate enlightenment. He likens the nervous system to an old telephone switchboard in a small town. He writes, "If the main telephone office wishes to stop communication with the different parts, it can turn off the main electrical switch and there will be no flow to the different quarters of the town.

"Similarly, the scientific method [of Yoga] teaches a process enabling us to draw to our central part – spine and brain – the life current distributed throughout the organs and other parts of our body. The process consists of magnetizing the spinal column and the brain, which contain the seven main centers, with the result that the distributed life electricity is drawn back to the original centers of discharge and is experienced in the form of light. In this state, the spiritual Self can consciously free itself from its bodily and mental distractions."

In this description, Yogananda may be describing a process of pulling back of power from the sympathetic nervous system to the central nervous system similar to what is described here as "pulling back your power".

Cleansing with the Infratonic S and Mag0$_7$ and Prayer

4:00 AM

The ongoing success of your endeavors is assured. Acceleration is brought about by prayer and commanding. Know that the strength of your will and intention are keys. Know that Mag 0$_7$ is clearing out your intestines, getting rid of old control behaviors and increasing the effectiveness of the Infratonic S on the limiting beliefs stored in the diaphragm and abdomen. Continue to use the Infratonic S for 20 minutes per day. You still have much limitation and preconceived belief to release. *[I have observed in my experimenting with the Infratonic S*

354

that using it causes increased elimination of sticky stuff from my intestines, and the Mag 0_7 is important to keep my bowels moving.]

You are a work in progress. When you are complete, 40 years from now, it will be time to transition. Until then <u>continue clearing, cleansing, praying, commanding, and intending daily.</u> This is so powerful.

Infratonic and the Vibration of Love

The Infratonic does truly awaken love wherever it is applied. It does induce wholeness based on love. It is, thus, a universal healer. Everywhere it is applied it radiates out to the surrounding environment, not just to the people to whom it is applied.

R.: Here is my technical description of these devices:

The Infratonic 9 device: The 2nd field highlighted in our model of the vital field of the body is the magnetic field, or the emotional body. This is where the Infratonic 9 appears to be most effective. The Infratonic 9 produces mechanical and magnetic stimulation with stochastic resonance (noise in the range of the EEG of the nervous system). It calms and normalizes excesses in the activity of the magnetic layer of the vital field surrounding the body. It also unravels blockages in this magnetic body, which releases energy that can catalyze healing. A big key to its effectiveness is that it produces unpredictable signals in the range of the nervous system, which dramatically increases its effectiveness.

The Infratonic S device: This device also works in the magnetic field of the body, producing coherent linear magnetic pulses that are much more effective than the Infratonic 9 at engaging and freeing stuck emotions from the Etheric Template. In particular, subconscious limiting beliefs are good examples of highly resilient distortions in the field. Limiting beliefs are beliefs we don't think about, like "I am not worthy" or "I make mistakes." They cause us to create a world in which these beliefs come true over and over again, limiting our full expression. They get heavily embedded in the etheric template, and are very hard to remove. Post-Traumatic Stress Disorder (PTSD) involves emotional reaction patterns which got heavily embedded in the etheric template, and play whenever triggered. Migraines function similarly. Mechanical shock and loud noise help make it easy for this Etheric Sludge to get embedded deeply into the etheric template.

The etheric template is an ideal healthy pattern for the physical body in the electric, magnetic, and mental fields that surround our bodies, and is what directs healing so that a cut or fracture can heal completely and the original pattern is restored. Our thoughts and emotions are made of the same magnetic and mental materials as the etheric template, but are not native to this etheric template. Limiting beliefs are powerful products of our thoughts and emotions that block us from a wide variety of achievements and capabilities. They are not true from the standpoint of the etheric template, and limit us in the expressions of our lives. They are distortions to the etheric template, which may cause a whole host of physical diseases. Thus, it appears the Infratonic S may have application in treating chronic illness, unleashing optimum performance by dissolving limiting beliefs, and deprogramming PTSD.

Just as images on a CRT monitor can get warped by a magnetic field, the etheric template can get warped by emotions, causing pain chronic illness and impeded healing.

The Principles behind the Infratonic S:

[I came across somebody who apparently had an extreme case of embedded limiting beliefs, and got these suggestions:]

The Scalene Light could have dissolved the mental structures holding this mess of astral/emotional structures in place. Nancy would then have been desperately trying to hold her life together as these limiting beliefs would painfully hemorrhage out of her. Thus the Scalene Light was determined to be an unstable, uncertain, and uncomfortable plan in this case. It would probably have worked, but was not an elegant solution.

We started with the Infratonic, which was helpful in softening the edges of this highly structured and highly self-protective emotional mass of restrictions. Then we brought in the hybrid, the Infratonic S with the help of her husband Robert. *[Robert started sending me articles about Phononic energy. I had known nothing about it, and as I read those articles, I realized that the Infratonic S was a strong possibility, creating planar magnetic phononic waves that would penetrate much more deeply than the Infratonic. So I built one.]* He wants her to be joyful and free even more than she does. Having his participation makes this treatment method more compelling, and ultimately, more successful for her.

So the Infratonic S has been dissolving the tangled mass slowly, steadily, without tearing apart her mental fabric as the Scalene Light would have done.

It is best for you to apply Infratonic S starting with the Aware setting followed by the Melt setting for the next few days for as many hours as is practical. Then, on Sunday, in the central square, you can apply the Scalene Light to cut the mental bonds that have been holding this mess together. She will need a warm salt bath and a deep sleep afterwards. She can then continue with the Infratonic S until Tuesday, when the Scalene can be repeated to finish the clearing of the deeper restrictions that were protected on Sunday.

This is not a treatment plan. It is a vision of one possible future. When you read this, events are likely to accelerate, so we ask you to

357

release all expectations of events, schedules, and procedures. *[She tried the Infratonic S for a few minutes, felt additional discomfort, and chose not to continue, so I can't say it was a success in this case.]*

Phononic Infratonic

The effects are subtle, yet profound. A coherent, planar magnetic wave propagating through the body has a very strong effect on the emotional body, much stronger than an expanding spherical wave. The planar magnetic wave can target particular stagnant locations and increase astral/emotional cooperation rapidly. Grief, fear, and anxiety will be swept away much more rapidly. Targeted joint healing. Almost immediate reduction in inflammation deep in the body. Definitely worth investigating.

Phononic laser provides coherent conduction of magnetic pulses through liquid crystal structures within the body. Given that the body's crystalline structures are generally amorphous, such a stream would have a different effect from a non-coherent signal. You already get coherent signals through bones and meridians independent of the phonon source. Powerful magnetic pulses are the product of this investigation. Experiment with hand emission.

The Infratonic S is excellent for emotional processing including:

1) Dissolving subliminal limiting beliefs
2) Dissolving emotional cell memory
3) Dissolving stuck emotions
4) Dissolving stored emotional trauma in tissue

The three signals are

Melt: This is most effective for dissolving subliminal limiting beliefs and other emotional sludge where there is no conscious connection to it.

Heart: Where there are heart issues associated with the residual emotional sludge to be dissolved, this signal is best.

Aware: This signal is best for bringing buried emotional issues to conscious awareness, and is best where it is desired to learn the conscious patterns behind the emotions.

Where the stuff to be melted is simply old sludge without conscious connection, or comes from others, or when you just don't want to get into where it comes from, start with Melt. Otherwise, start with one of the other two signals. Where you are looking to learn karmic or past life lessons or something like that, *Aware* is the best place to start, and where the issues are stuck in and around the heart and lungs, or around heart issues, then start with Heart. It is often best to finish with Melt to clear away left-over residue, and to follow with the Infratonic or Mobile Medic for calming and smoothing.

Infratonic S for Nancy

[I dreamt of cleaning up the kitchen that had been flooded. I pulled weeds out from the walls. From one corner, I managed to pull out a bundle of weeds about 6 inches in diameter and 6 to 8 feet long all together, and flexible at 3 bands. Later I learned that I had rescued my twin brother, who had been cloned from me after I went into enemy territory (Chinese) to rescue my love from a pond of underwater bamboo weeds where strange war research was being conducted.]

Good morning! You slept deep and long. The Infratonic S is doing very well in dislodging long-standing limiting beliefs. It is working well for Nancy. She does feel it strongly, but has not gotten her head around the possibility that she could be freed from what she considers to be herself. She identifies with her unconscious limiting beliefs as you do yours. The difference is that you trust me. You are willing to try out my suggestions and willing to suspend your beliefs because so much has already changed. You are asking her to suspend all those limiting beliefs which she has held for her entire life, and has thought to be a part of her.

She does not know how to let go. She has lived her life holding it all together for herself and her family. She feels her strict structure of code of ethics has served her well and is a key to her survival. This code has been so strong and rigid that, at some point, her body started searching for a way to free itself from this restrictive code. The infection was partially effective, in that she now perceives the code as something she needs to get rid of.

The problem is that one side of her wants to get rid of it and the other part of her feels she will die, that she will lose her identity/power/authority if she lets go of it. A part of her will die. The part which is restrictive.

Making the Etheric Template Transparent

Here's another method for clearing the etheric template: As you lay in bed, just before you begin to fall asleep, pull back your power from all fears and self-limiting beliefs. Intuitively find them throughout your body and field. Pull back your power from all.

This offers a broader clearing. You don't need any fears, limiting beliefs, or residual thoughts or emotions. You are guided through this process. You don't need your memory constrained by this means of residual storage. Everything in your etheric field can be pulled out by the use of your own intention to do so. Another way to look at it is to make only your etheric template transparent and pull your power from all that is left, including all debris embedded in the etheric template.

You have access to perfect memory of past and future as required. All residual is noise. You are able to pull back your power and let it all dissolve at will. You will awaken feeling Joy, Love, and Bliss.

Expect to be filled every minute with Joy, Love, and Bliss. Feel it now, in this very moment. Soak it up and breathe it in now as you visualize it flowing all around you. Know that you are blessed and that you are a different person now, with a lot more courage, insight, and passion. There are no limits to what you can and will achieve.

Remember, as you fall asleep, to 'Pull Back Your Power,' from all fears and self-limiting beliefs. Find them throughout your entire being and energetic field. Employ a collective draw to pull back your power from all.

Joy of Anticipation

Joy of anticipation is an experience unique to Earth and other time/space environments like it. Anticipation is a most interesting phenomenon, constructed of expectation, which is a built-up mental structure regarding future events. It can be called illusion, and it can be called a step toward creation. Anticipation can be a powerful creative tool and it can be a painful disappointment. The key ingredients here are clarity of intention and, the most difficult one, releasing the outcome. How can you believe over and over that something will happen when it hasn't happened yet? A big part of it is trust, surrender, releasing the outcome, and expanding possibilities.

Another huge part of it is clarifying the vehicle 'You'; understanding the creative process and all of its components within the vehicle of creation. Are your actions and intentions consistent and able to bring about the anticipated outcome? The anticipated and desired outcome eludes you. You hesitate getting started, and day after day, week after week, you question, "What am I doing wrong? What could I be doing better or differently?"

Release your beliefs and expectations daily, every minute, and see who you become. Remember that the central alchemical process is 'You.' You may think that you are honing the world, but it's not about the world. We are honing you. How can you let go of your beliefs and expectations and still be becoming more every day?

The more you release striving and surrender to the emergent moment, the more rapidly you become more abundant. Becoming more is about releasing layer upon layer of limiting beliefs. You are a bright light – a brighter and brighter light, which is destined to fill the entire Universe. That's it! All of your beliefs and expectations are just manifestations of the prison walls that keep you from shining brighter.

Perceive the newness of every moment. Release the filters of expectation that cause you to see future expectations rather than simply the joyous emergence that is in front of you. Notice how you make many emergences into emergencies by comparing to expectations. Live life in the Now; free from expectations, and experience the expansion of your light every minute.

V. Descending

27. Rejuvenation

Many years ago, I read "The Emerald Tablets of Thoth the Atlantean" by Thoth the Atlantean. I would read it over and over, entranced by the rhythms of the poems. Over the years, I would pick it up again, and each time, it felt as alive as the first time I read it, though no more comprehensible. I was most intrigued by the Halls of Amenti and the Rite of Rejuvenation, described as practiced by Thoth to maintain his physical body for hundreds or thousands of years. I wondered what it meant.

About a year ago, my lifetime research partner, Maria José, (who loves her name, claiming it means "mother and father of Jesus"), told me of her dream to build a rejuvenation chamber which would reverse aging through removal of a lifetime of congestion. She proved to be quite a dreamer or visionary because she also expressed a desire to dissolve cancer through vibrational frequency devices along with a desire to master methods of teleportation and time travel, and other exotic aspirations I found inspiring. We developed plans for a rejuvenation chamber and rudimentary teleportation technology, but have not yet built prototypes, partly for lack of time, and partly because I lacked the confidence that they would work. I know I must build them someday, just in case they do work, and also because Maria José occasionally prods me onward.

Recently I met a young Ecuadorian shaman who grew up in Vilcabamba and who was familiar with the traditional ways. He explained that one method used by these centenarians to prolong their lives was doing a San Pedro ceremony every 3 or 4 months during the full moon. The idea of clearing out old patterns of congestion from the body with the use of the light or energy of the full moon was common among all of these seemingly diverse rituals. Also, the goal of clearing out congestion is what led me to develop the Scalene Light and Infratonic S as methods to remove Etheric Sludge, and further, to write this book.

I was quite surprised earlier this year, when I came across a channeled book which claimed to reveal some of these secrets. I read "Anna, Grandmother of Jesus" by Claire Heartsong, in which Anna claimed to have lived for more than 600 years. Early in the book, she said she would describe the method she used for her Rite of Rejuvenation later in the book. I read each chapter with great anticipation that her method would be revealed. By the end of the book I was disappointed. I had not found the secret revealed.

However, I asked during my early morning writings about this topic and wrote many interesting things, among them, that the elderly who lived in Vilcabamba to be well beyond 100 years old did so, to a large extent, because they lived in an isolated community which believed that being healthy and functioning at a high level of health beyond 100 years old was not only normal, but real because several members of the community were, in fact, 110 and 120 years old. The fact of a healthy trans-centenarian life helped them to believe that it was possible. Researchers in the 1970's found these centenarians working in their fields or sitting under trees reading books without glasses!

Below is the result of one of my early morning writings, which describes a method of rejuvenation. My scientific curiosity compels me to research this one! I will not know whether it really works until I have tested it fully over the next 100 years:

Purification and Life Extension

4:00 AM

Any discomfort you feel is Etheric Sludge seeking to escape. When you feel this, always take some time to drain it out. Daily is very good. As you do, you will become more aware of feeling clean and of feeling a bunch of free mobile Etheric Sludge in your body calling to escape. You will find that standing with your hands and feet open when stuff wants to escape is exquisite. It is bliss as the sludge flows and escapes from your body. You will find it is easiest when you put the Scalene Light on the floor facing up, first on clear, then on soothe, a few feet away from you.

Plus the Scalene Light cleanses the Earth when you are done. The full moon offers an added impetus to expel the Etheric Sludge. Essentially, standing with the full moon overhead, or anywhere more than 30% above the horizon, will provide a downward fluid wash of lunar etheric water. 10 minutes with the Scalene Light pointing up nearby, standing barefooted on earth, is sufficient to release all the Etheric Sludge in your body provided you can maintain focused intention to do so for the entire 10 minutes. It feels so good that you will feel called to do it every month.

Will doing it on wool carpet on the wood floor indoors work? It can work. However, not at first. You must do it a few times outdoors, on earth with bare feet. Also, be sure to call in a sacred space with guides, teachers, Angels, and me. We will guide you. You will do it, succeed in deep cleansing, feel it happening, and know it is real. You will feel how good you feel during and after, so you will want more.

Without the Scalene Light, you will find that there are some really thick and hard chunks of Etheric Sludge that get stuck on the way out, in shoulders/hips, elbows/knees, wrists/ankles, or hands/feet. This can be painful and frustrating. The Scalene Light will help. Also, if needed, the Infratonic S will soften the sludge and lubricate the channels, freeing up the most transigent sludge.

First, do this clearing in a standing position, then lay down with your knees bent and feet flat on the ground at the time of the full moon. Your goal is complete freedom from limiting beliefs. You have done it before. It is easy for you. Practice a few times during the day [before the full moon rises – for practice]. Soil works better than bricks for your feet. Your head faces the moon and your feet face the Earth. Pad is fine. Your intention is that you let go and allow the current from the moon to wash through you and into the earth.

You will feel as if you are swimming in the water, feeling the flow and expansion. You will love the world of energy you discover. Just float and trust and you will be delivered. No need to plan anything. Feel yourself being held in the water. Know that you are safe. Observe all the interesting things you do, and know that it is the real you doing them.

Yes, do a Meetup during this full moon and try it out. Three or six people at a time dancing around the Scalene Light and around the fire will work well, but follow-up by doing it alone the next night because silence, solitude, and patience will release deeper sludge. Remember, you need 10 minutes of continued focused attention and intention to achieve the cleansing. In addition, you shall find that, as your body is learning to do this deep cleansing, it will work best if you do a gross cleansing (wash cycle) the first night and deep cleansing (rinse cycle) the next night. As you become familiar with the process, you will find that one night a month is adequate.

The time for eternal life has arrived for those who choose and are chosen. It is not what you think. It is an ever growing community with Heaven on Earth. It is a level of joy, love, and bliss that can't be known while living in a sea of sludge. This is a new experience. Don't expect it to be at all familiar, or even to recognize it as it emerges. You shall know it by what it is not.

While you have come to be aware of a world of familiarity, experienced through the 5 senses, through the filters of expectation and limiting beliefs, you have no idea what the 5 senses feel like with clear, unfettered filters. What's more, each of the 5 senses speaks to each of the 3 worlds of human endeavor, or in this case, human experience. Thus, prepare for 15 flavors of ice cream, 5 flavors of mental joy, 5 flavors of heartfelt love, and 5 flavors of powerful, physical bliss.

As you get really clear, calm, and sensitive, years from now, and have cultivated very potent focused attention, you shall experience all 15 flavors at once, playing a symphony of communion with Heaven, Earth, and Nature. This experience is so profound that you shall want it all the time, and shall make every effort to clear Etheric Sludge on every corner of the Earth. As the sludge of the world clears, the symphony shall

366

become more exquisite and resonant to experience for all who choose this pathway.

The call goes out to all lightworkers, and all on this Earth at this time are lightworkers, mostly mired in Etheric Sludge. Darkness is the density of light. Expose it to the light of the Sun and the Moon, and even the darkest of sludge will radiate brilliant loving, healing light that will bless the Earth as it expands to full brilliance.

It is enough to bring your own darkness to light for you are one with the All. Bless any darkness you perceive in others, knowing that it is simply illuminating light, stored and concentrated for another day. Reach deep inside yourself for even deeper, finer layers of Etheric Sludge, bringing it all to light until you are radiating with the light of the Sun.

Know that you now live upon a New Earth and that the dregs of the old world are quickly and radiantly dissolving. Be patient and take joy in knowing that all is evolving perfectly. There is no need to do anything besides reaching deep inside and bringing all the sludge you can find to the light of day, and living fully in the symphony of that light.

Longevity: The Reversing of Aging

[The concepts described in this section are a stretch for me. I can certainly believe that healthy diet inevitably increases lifespan, that exercise, fresh air, clean water, and fresh food are part of the formula. I can certainly believe that embedded Etheric Sludge in the form of limiting beliefs limits the length of people's lives and causes chronic illness, where health and vitality are our true nature. However, the concept that eternal life on Earth in a human body is an available choice is quite a stretch for me. Nonetheless I will persevere and see what happens. Already I feel younger, healthier, stronger, faster, more intelligent, and more comfortable in my body than I was 10 or 20 years ago. For those of you who read this section and find that you know it is true, I say "amazing." For those of you who read this section and assume eternal life is impossible, I suggest you join me in an attempt to

find limiting beliefs in our bodies and remove them so that we can consider this as a possibility without preconceived judgments. Here we go:]

4:00 AM

Your method for treating people with cortisol consciousness is similar to that used in the reverse of aging, and is similar to the San Pedro cactus ceremony. Clearing of the body from hidden residual thought forms stops aging. Conducted consciously, and monthly, this method sustains the body at a young age. [*Cortisol consciousness refers to a physiological state in which a person's conscious awareness is dominated by fears, and their blood stream carries lots of cortisol, a hormone which increases vigilance.]*

The word "consciously" is a huge word here. One must become aware that residual thoughts and emotions, and particularly the belief that a lifetime is less than 100 years, are the beliefs that cause aging. One must come to experience these limiting beliefs as external to oneself, and removable from the etheric template. When this awareness is achieved the elimination of these limiting beliefs (of which fear is one) becomes easier.

San Pedro ceremony, when conducted consciously and in communion with people who understand this method of rejuvenation, allows new sustaining beliefs to be implanted at each rebirth. Just as limiting beliefs can be implanted, woven into the etheric template, unlimiting or liberating beliefs can be woven in, which will inhibit the entry of limiting beliefs. All seeds of chronic illness are limiting beliefs that can be eliminated. The Essene community provides a source of such beliefs, as does a community with Spirit and community with Nature.

It is hard to be first, or to feel like you are first. It is much easier to have a community like that in Vilcabamba 50 years ago, or with the Essenes, because living examples make the possibility of eternal life more credible, and it must be 100% credible.

Subluminal limiting beliefs are so deeply woven into the etheric template in recent times that a method such as the Scalene Light, Infratonic S, or San Pedro ceremony is necessary to untangle the etheric template. If people are to achieve this aim, they may conduct an etheric template cleansing as a ceremony with the specific intent and with strong resolve to provide youth, abundance, and infinite health.

A knowledge that you are Spiritual beings with physical bodies, as opposed to either physical beings with some external physical presence, or physical beings alone in an unconscious universe, is necessary to be able to achieve eternal life of your physical body. The whole reason that physical bodies must die and be recycled is because these limiting beliefs get so thick in a physical body that it cannot survive, just as nematodes build up at the roots of tomato plants, apparently killing them.

Etheric Sludge has grown so thick in recent centuries that very clear, powerful rituals must be conducted regularly with strong resolve to clear it out of the body. This is because, as soon as it is cleared, other Etheric Sludge begins to enter from the environment and replace it, imparting a high degree of numbness and disconnection from Spirit.

Thus to achieve or restore the Garden of Eden, Etheric Sludge must be cleared from Earth. Two methods can be employed simultaneously. First the environment can be gradually cleared through regular ritual using powerful methods. A meditation with clear intent can be employed, though clear intent is difficult to maintain over the extended time necessary, precisely because of the Etheric Sludge. Thus devices such as your Ra-Ad are preferred at these early stages of terraforming. The second method is teaching large self-sustaining groups of people that reestablishing the Garden of Eden and living lives of eternal abundance are practical goals that are easily achieved. With persuasive, pervasive teaching, this can and will be achieved. You must not only lead with tools, but by example. Not only by extending your own life, but by teaching, guiding, and inspiring thousands of others to do so.

The pattern of longevity observed in Vilcabamba started with a single person who, through ceremony, meditation and clear intent, along with Spiritual communion, was able to reverse aging and live an extended life. The long-lived people of Vilcabamba all, or mostly, died because the belief patterns of the invading researchers in the 1960's and 1970's were so strong and determined. They penetrated the population of centenarians and embedded powerful limiting beliefs into them that they "should be dead by now." The community of 15 centenarians was sufficiently strong and coherent that they could hold their own beliefs in an insular town of 700, but when Vilcabamba was opened to a world of billions, the small number of centenarians was overwhelmed. Essentially, the placebo effect, the belief that illness and death are inevitable without doctors and pills, was embedded in these centenarians.

The Spiritual communion was weak. It must be strong. The environment was full of Etheric Sludge. It must be cleansed. The individual bodies must be cleansed every month, thoroughly and lovingly. The turning of the ages, the cleansing rays of the new Sun, the rise of spiritual awareness among much of humanity, all make this task achievable. It is of a high priority, not because longer life of physical humans on Earth is desirable, but rather, because a world of sustainable abundance can and must be achieved as a part of the new emergence of Gaia. Do your part and others will do theirs. Know that it is done, and that you are held by spirit every step of the way.

28. A New World, A New You

Truth Shall Emerge Triumphant

4:00 AM

The New Earth sings, and everyone loves the song. Truth shall emerge triumphant because it is noticed, loved, and chosen. You will observe this emergence everywhere you look. The old will be crumbling and nobody will notice. The new will be emerging and all are resonant with its Bliss. Feel into this amazing dance every day, every minute. All aspects of Earth were reborn 4 months ago *[12/21/12]* and are rapidly metamorphosing. The beauty of this budding flower is so dazzlingly beautiful that it can't help but be noticed. It cannot be named from the past. It is simply the new emergence of Truth.

Watch for the out of place, the unexpected, the spontaneous, the serendipitous. This is the hallmark of the new emergence. Watch for it in the words chosen and not chosen, the topics skipped, the eye contact, the food presentation, the chair arrangement, seating arrangement. Every detail is different than it was 4 months ago. It is not what is out of place. It is what fits so amazingly perfectly.

Just as the dawn is a sure sign that the light of day shall be triumphant, the blissful vibration you feel is the sure sign that the New Earth has arrived and is unstoppable. Bless the Bliss of it all!

Truth and the New Earth

The Scalene Light makes everyone who is living in their own illusions a little uncomfortable, depending on how attached they are to their illusions. This is its principal, if only a side effect. This is the same effect as "dissolving armoring." Sometimes the armoring or unconscious illusions are so deep and unconscious that people won't even know why they are upset by the Scalene Light.

Illusion has no place in the New World. Truth is that which is in alignment with the New Light Grid that is descending on the planet. The reason it is slower to descend is because there is no place for it to descend, as long as a person's thoughts, their belief field, is occupied with illusions. Once the field is cleared, the new grid can be received and perceived. People learn new thoughts, ideas, and truths more quickly once they are cleared with the Scalene Light. Replacing illusions with truth, or replacing incompatible beliefs with coherent beliefs, makes all aspects of life easier. The only hard part is letting go of those illusions, those incompatible beliefs we want to keep.

One of the blind spots of many who follow the new Light is Trust. It is also an effective tool for disengagement from the illusion of others, allowing these others to follow their own paths to limitation and eventual disillusionment, opening the Light of Truth. This Trust involves trusting that the pathway of illusion is slow, and often painful, but sure and true. This is a good new-paradigm answer to the question of why God allows pain and suffering.

The joy of life is not an elusive quantity. It is what is left when people let go of illusions of past identification and future desires, and settle into the moment, being ever guided by the ever evolving and changing Truth.

You can shine Light and dissolve darkness, but this darkness also has its truths. Darkness must unravel for each individual in a unique way, which is optimal in the moment. Your work of dissolving illusion or Etheric Sludge on a global scale makes life and evolution easier for many, but does not "relieve suffering." People choose to sit in, or release, their seeds of suffering by choice.

This is why Earth is so prized around the galaxy. There aren't many places in the Universe that allow us to muddle in illusion. Maya, Satan, and Lucifer are teachers for those who want these pathways and Earth is a remarkable gift for those who choose pathways of illusion. And, what truly remarkable and believable illusions you are able to create on Earth!

As with all things, the old Earth has ended and is disintegrating. The New Earth is ascending very rapidly now. Relax, float in this river of change and feel the textures as they float past you: Joy, pain, and sorrow. It's up to the individual to experience this ever-changing flow with their own framework of unconscious beliefs, desires, aversions, and traumas. For you, float on the river and experience the bittersweet yet blissful emergence of the New Dawn.

Blessed be all of God's creations, emanations, and manifestations. Blessed be!

The Emotional Body: An Alien's Perspective

Quantum Mechanics: Space and time are illusions. In the real Universe, time does not exist. The past, present and future exist together in the eternal now.

Beings who live on the other side of the barrier, where we are all one and creation happens instantly with the snap of a finger, find that creating a home or a car is effortless. These beings may tend to get cocky, ignoring all the problems they bring about by creating things for themselves. They just don't understand the responsibility of co-creating, and have no idea of what's involved in the process of creation. Earth was created for beings like these, who want to get in touch with the details of this process.

While they conceive of an idea, a complete creation with one thought, it has all been slowed down for us. In fact, the creation of time and space means that these tightly packed 'ideas' are incomprehensible in our 3D world. Thought has to be linear to fit into our brains, so we can deal with only one small concept at a time. We need to put a whole bunch of small concepts together to get anywhere near to what they consider an 'idea.' Then, we need to put our plan into action which, as we all know, requires a very strong desire to continue moving forward and to overcoming discouragement and obstacles. Then we need to start building. It often seems that we should be able to get it done within a few minutes or hours, but it winds up taking days or weeks. And if we get discouraged, our plans may go on the back burner for years.

These 'Gods' and 'Goddesses' don't understand this key part of the creation process, the emotional level. Without the desire to get it done, we don't succeed at anything. And this magnetic emotional plane is so full of turbulence in the form of angry and depressed people and conflicts with other people's plans that it's amazing that anything gets done at all.

There's another challenge that the 'Gods' and 'Goddesses' simply don't understand. This emotional plane, this field of spinning and swirling magnetic substance, is full of places to get lost. This is the astral plane, where nearly all of what we encounter is an illusion. It's so easy to get full of anger, grief, depression, or fear, which isn't present in the relationships between people or in projects. It exists only within ourselves and is an illusion. We often get swallowed by the astral plane itself, living lives of quiet desperation, or living out a childhood of anger, control, or victimhood over and over again.

The 'Gods' and 'Goddesses' have no idea how hard life on our side of the space time continuum is. Only by living through this Earth plane existence do we begin to understand what true cooperation is. We eventually learn the difference between creation from the standpoint of separateness and cooperative creation where the wellbeing of All is considered. Earth is a mess. The Gods would probably just snap their fingers and erase what we call progress.

Raising Up the Energies from Institutions and Legends

[Written before 12/21/12] The principal process of transmutation proceeding on Earth at this time might be described as raising up the energies. An aspect of the etheric of Earth is lifted, and if we are open, we are lifted with it. A wave of finer energies then washes through us.

At this higher level, the dross drops or washes out of us. The atmosphere is so filled with this dross, discarded thought forms, that it is pretty chaotic. We will never return to the past, and can't conceive of the future. The now we live in is one of chaotic change.

The only reason we are not uplifted with the energies as they rise is because, at a deep level, we are holding onto the mind-thought of institutions and legends. This attachment to the past causes us to perceive change as evil. Institutions that are feeding this fear will crumble because they are not based upon Truth.

When we let go of these mind thoughts of institutions and legends, we naturally start to be lifted. This is not easy because our whole way of life is based around institutions and legends. I'm being shown a picture, a huge rock, alive and unattractive. If these mind thoughts open, we are freed to be lifted. Attachment to any traditions might be holding us back. Affiliation with political parties, religion, corporations, family, media, diet, transportation, and use of energy will all change.

Lifting or raising your vibration is not what you understand. It's not letting go. It's so much more. It's like the body becoming air or as light as a feather. The denseness (belief structure over thousands of years) will simply not be there. You can't even imagine how it will be because there has been nothing like it. Imagine feeling the true heart that feels so other-worldly and living in it all the time. It can't happen now because it is too fine an energy to resonate in the body now.

How can we become more comfortable during this process of lifting? By rising above these old patterns, the discomforts will fall away.

The New Sun and Intelligent Materials

The new Sun of your solar system chooses to shine with a new light, a light which dissolves certain aspects of matter as you have come to know it. What you call physical consciousness is an example.

This is difficult to explain through a body which has considered itself to be "physical" for many millennia. This plane of existence is now phasing into a nonphysical state. Nothing disappears. It is simply that this material becomes transparent, or unperceivable to the human sensory organs.

Those who have much to work to do on that plane of reality will be able to do so, but will need to transition to another plane of perception. They will choose to reincarnate elsewhere. This will be perceived as a lower birthrate around the world, in many cases by choice, whether individual or societal, and in other cases, by low fertility. This will take many generations. It will become accepted that a lower population on Earth is desirable and is achieved by human choice.

Phasing of the solar light will also bring a new light, a light of profound love, to the sensory organs of the human body. This means that you, the human race, are becoming conscious on the level of the circulatory system. This is a whole new mode of thinking and experiencing. You will find the shifts are so fast, so radical, that governments, religions and other organizations held together with logical constructions will become not only obsolete, unsatisfying because they no longer have substance/foundation, but also because new social structures will be so brilliant and inviting that they will be quickly adopted. The world shall be governed and organized based on love rather than logic.

Know that you have much shifting to do. However, you have made the decision to be flexible, to let go of all that is dissolving anyway.

Science will shift radically as old ways of perception and measurement become discarded and new "superior" methods are adopted. This will lead to a whole new means of creating, which, to a large extent, will replace what you call "manufacturing." Synthesizing, materializing, and accreting shall be better descriptions of the methods employed. Such materials as steel and crystal shall be understood as conscious beings or existences, which can be invited, through intention and prayer, to come into existence. There will be many principles of accretion in which materials, prayers, and intentions are brought together. With this, products, living, loving, communicating products, will spring spontaneously into existence. They shall have long life, qualities of self-repair, and the propensity to teach owners and users how to get the most out of them.

They shall not be possessions as that is an energy of the past. They shall be friends and companions. A spaceship will teach its occupants how to use it and a myriad of other things. Teaching shall be "in the moment." No facts, nothing to remember. Simply a knowing that this is the way it is always been.

A human jumping from 30 years ago to 30 years in the future would have difficulty. Intuitive interfaces, built in user's guides, children being able to learn seemingly complex operating strategies overnight, will baffle the older generations. They must let go of all preconceived beliefs, truths and assumptions.

Sunspots

You are willing to let go of old beliefs that no longer serve you. Your daily application of the Scalene Light takes off the surface layers so unconscious belief patterns can more easily come to the surface. Think of it as decompression. The sunspots are an outward sign that the Sun is active this year cleansing His solar system of the old ways, to make way for the new.

Yes this is generally applicable. Those who are applying a method of field cleansing with meditation, introspection, affirmations, focused on letting go of old unconscious beliefs, are sleeping much better than those who don't.

Eating veggies provides a better night's rest then eating meat. Wild freshwater and deep-sea fish and free range chicken disturb sleep much less than "conventionally grown" meats.

The Infratonic also clears the field and contributes to better sleeping. Now, why the deep sleep? The solar flares, if you are relaxed and willing, wash away your field and all the worries and wants and fears you have programmed into it. This leaves you feeling unmotivated, either by beliefs or by desires and fears.

It can be helpful during sunspot times to write your affirmations, goals, and intentions in the evening, then to read them in the morning when you're ready for the day. Those who drink coffee or tea can read them while drinking.

Those who believe they "should" keep the "edge" of unconscious buried motivators will try to hang onto them, to protect their fields from the sunspots. This can cause a tightening up of the field and a sense of stress all day long and disturbed sleep at night.

Another effect of sunspots, or regular use of the Scalene Light, is the gradual increased clarity of perception of unconscious control dynamics, often called "entities" or "demons" by healers and religious groups. These are buried belief, desire and fear structures that we, at some level, believe we need. However when we perceive them clearly, we realize they not only don't serve us, but make a shambles of our lives. As we examine them more deeply we begin to perceive them as "not us" and we become open to the notion that they are entities or demons.

In fact, they are not us and are controlling us, but so are fast food jingles, fear-based media programming, childhood training, and traumatic incidents.

When we get down to it, none of it is us. This is the secret behind all great religious and successful mental and emotional release techniques. None of the unconscious beliefs, desires, and fears are us. They are nothing more than unconscious reactive patterns. If we perceived our lives more clearly, we would realize how much we seem to be possessed by demons, and entities, whether consumerism, polemic politics or eating habits. None of it is truly us.

Solar flares are a blessing. Like the rain, they wash away the old and make way for the new. Regular use of the Infratonic at thymus, solar plexus, crown, occiput, perineum, or bottoms of the feet, will soften the desires and fears, making life easier.

Solar flares are a way in which off-worldly influences shape our lives. The ways in which Earth affects our lives would seem to be more influential, yet this is not the case. There are powerful forces like tornadoes earthquakes and volcanoes, but none of them have the potential for influencing all life on Earth as much as solar influences.

It is also true that the Central Sun or a black hole has an even greater influence on life on Earth. Fortunately for you these changes are generally gradual and last a long time. One important example is that of the oscilloscope. It measures lots of vibrating frequencies that originate from the Galactic Center.

Richard and the Galactic Center

The sound in your head originates at the Galactic Center. In important ways, it is you. Your life is at the Galactic level. Be aware of this level of life and of its influence on your life. The illusions of life on Earth are orchestrated and agreed to at the Galactic Center. You still say so what? You are stubborn this morning.

The whole world you know is orchestrated at the level of the Galactic Center. This is huge. You only slightly live your life on Earth. Mostly, you live and experience life in a mental world of your creation.

[That was all I the writing could handle this morning. It continued the next morning...]

The Galactic Center

Yesterday morning you were unable to handle the very high frequency of the Galactic Center. It is your home, your life, but your body is still of sufficiently low frequency that sustaining Galactic Center frequencies in full consciousness is a major challenge. With your clean diet you will soon be able to handle it.

Cooking food reduces its frequency. Yes, the pie. It was the lowest frequency food you had yesterday and you had a lot. It's a lot higher frequency than chicken or fish, but still not quite high enough even for Pleiadian consciousness.

Your *LIFE* at the Galactic Center is full of love, full of joy, full of mystery, learning, and surprises. You love it there and have a tough time with incarnation on Earth because it is so slow and painful. You remain fully conscious at the level of the Galactic Center. It's just that your body on Earth slows down so much that there is just a dim awareness of your *LIFE*.

With the new Sun, which is set and determined to dissolve the remaining slow vibrations on Earth, awareness of your Life is becoming easier and easier.

Raising your Vibrational Frequency

The Scalene Light is a two-edged sword. On the one hand, you carry a strong determination to break through the barrier to your LIFE, a determination which would be mostly erased by the Scalene, at least temporarily. On the other hand clearing with the Scalene Light clears out Etheric Sludge, which is a major obstacle to become conscious of your *LIFE*. To benefit most from the Scalene Light it is best to reset your intention after cleansing with it. It is helpful to write down your intention, then after cleansing, read it out loud. This will help you to achieve whatever goals you choose.

Food: Eating Vilcabamba foods raises frequency rapidly, making conscious awareness of your *LIFE* much easier, as does eating fresh foods from your father's garden. Also praying that the low vibrations of your food be dissipated before you eat it is very helpful. Holding your hands over your food with this intention is especially helpful. A retreat in Vilcabamba will help you and others to approach your true lives consciously.

Infratonic S: How can I get to that high-frequency now? Definitely use the new model of the Scalene Light in the future for this cleansing. Your clearing of your upper spine was particularly effective.

Also, you intended to remain awake without dwelling on it, which was great.

The problem remains your solar plexus. We would like you to use the Infratonic S a couple of times a day over the next several days on your solar plexus, diaphragm area to permanently clear out some very old and relatively new low-frequency vibrational patterns. Leave it in place for a minute or so at each location for 10 to 20 minutes total.

Yes, your elimination – smelly elimination – was prompted by use of the Infratonic S on your solar plexus. Expect more such purging depending on your diet.

Home in the Galactic Core

The Galactic Core is your true home for this epoch or so. Like so many others who walk the Earth today, you have chosen to be clothed in a human body to teach and exemplify higher values in a world mired in millennia of Etheric Sludge. Also, like so many others, your nervous system consciousness had no more than a glimmer of this awareness until very recently.

As a Presence, a member, a part of the community of the Galactic Core you are, at your core, a powerful creator. However the rules by which Earth was created to function, and by the collective agreement of all who occupy Earth at present, you are not permitted to exercise this power to change anything about Earth except as you can manage through the very limited physical (nervous system) consciousness of a living human being, and from within the soup of Etheric Sludge on the planet.

There is not much point in describing *LIFE* in the Galactic Core. Your nervous system will never get it and the rest of you knows it fully. Our suggestion is that you simply accept the knowledge that you live in both worlds, and proceed accordingly. You can live the *LIFE* of the Galactic Core within your own personal world on Earth provided you do not infringe on the rights of others to live in their own private worlds of conscious choosing.

Know and trust that each person on Earth is guided toward a mix of ecstasy and misery that is perfect for their own development, and the marvelous cesspool of Etheric Sludge each has chosen to live in, and from which each is destined to emerge triumphant, whether a few years from now, a few lifetimes from now, or, and this is possible for everyone who has managed to read this far in your book, to emerge triumphant this instant, through nothing more than a firm resolve and a conscious choice to do so *NOW*.

Okay, now open to the vibration of the Galactic Core. The sounds and vibrations in your head get louder. You become less focused on the present. Feel your LIFE at the Galactic Core.

It was the alignment at December 21, 2012 and June 21, 2013 with the Sun, Earth, and Galactic Core that were awakening for you. Feel the vibration flowing into your body. Enjoy the feelings of strength and confidence.

Your concept of an individuated being standing on the Galactic Core is limiting you. Just be with the energy, the feeling. Notice all the jumble of thoughts as you begin to slide with the vibration. Following this vibration while continuing to write is a big challenge now, but will become easier with practice. Sit with it now. Soak into it and feel the heavy energies dissipating or draining out.

Know that your life on Earth is just beginning. You have been a caterpillar awaiting emergence.

Sleep now with the intention of feeling the high vibration swallowing you and the lower vibrations flowing and draining away, dissolving in the Light of the New Sun.

"I am the Power and Presence of GOD, now experiencing myself as whole and complete, having pulled my power back from all illusions of suffering. I am full and complete. I am forgiven. I am the world around me, and I love me fully and completely. I am forgiven! I am joy! I am love!"

29. Hydronizer...It Works!
...But Not Quite Yet

"One human who holds clear, unwavering intention, humble prayer, and firm command, has the entire universe behind him, and cannot be denied."

I have built many devices over the years. Some, like unlimited power generators, did not work. Others worked spectacularly, but I just let them sit. Some, I know how to build but have let sit, waiting for a day with more time and resources. A few are in production. Some of the devices I was most convinced would work did not. This chapter is about those creations which have not yet made it, yet hold great promise. I will let my writings tell most of this story.

4:00 AM

A beautiful morning. Full moon opportunity. Great morning to write about something new and interesting: For many years you have been interested in why things work and how they work. You have now discovered that they work because Great Beings have decided to bless these constructs with their collaborative participation. Traditionally, once a "device" works, "scientists" study every detail of it and optimize every design feature. Under this new understanding, this optimizing makes it extra appealing for the Great Beings Who are paying attention. Science calls these design parameters "laws of Nature." We might say they are "God's preferences."

With your Hydronizer, you are in the process of teaching God new ways of doing things. The potential is there, has always been there, but the intentional construct has been lacking. Thus, for the Hydronizer to work, you must see it working and believe it will work. Over time, and with clear intent on your part, God will learn new ways of fulfilling your command.

We have examined the Hydronizer from every angle and have determined that it will work and must work. It is in the best interests of Earth and humanity, which is a plus, but not enough. Another factor for you to consider is that it already works! In the future, we sit here watching it, and it works great. The model you have constructed is sitting in a museum in operation. It is so basic that young children understand how it works. They can feel and control spherical pulsations with their minds. They observe it working and, in so doing, strengthen the "principle" upon which it works.

For you see... God, that Great Being who blesses your Hydronizer, is none other than yours truly, you/Me. Remember that, without an earthling body, I have no authority to act, to create on Earth. It is only the physical bodies that get to "vote" on the laws of Nature. Thomas Edison had to build 10,000 light bulbs before he got one to work because he was building new operating principles into Nature.

The challenge, to a large extent, is twofold. First there is a noise of human beliefs which tends to make all those things which are not in the books of science "impossible", a faith that, if it is not already in existence, it doesn't work. The second kind of human noise is a thick soup of unknown sci-fi possibilities, an "anything is possible" attitude without any set structure. While this second form of human intention is what we have to work with, it is not very useful because it is so chaotic and filled with confused gobbledy goop, what you would call Etheric Sludge.

Hence, you now have the background to understand why you need drumming, Ra-Ad, etc., to clear away the mental dross to make way for the new. You also see the importance of the crystal grid, spirals, and offerings of water from the future, offerings of spherical resonance discs, and all the discussions of volcanoes and thunderstorms. You need to clear the field of old, chaotic thought forms, and renew your intention for an operable Hydronizer several times each day. You must intend and believe this so much that your entire garden, the entire neighborhood for miles around, knows that this is a reality.

The reason the Hydronizer is running at low power this morning is that it is engaging in subliminal programming, or imprinting of the thought forms of southern Orange County with spherical resonance and the possibilities that it brings to Earth. The Hydronizer is the first ever large-scale spherical resonator on Earth. You have built them elsewhere, but never before on Earth. It would not have worked before 12/21/12 because the "old paradigm" was too strong. As the old world crumbles there is a huge amount of fertile chaos on the mental plane to build anew.

Anything is literally possible now. This is why your head is so full of new ideas. People around the world are bringing new collaborations into existence. This time, after the shift, is the pinnacle of alchemical creative potential. Your Ra-Ad, your garden, protective crystal grids, and the physical structure of the Hydronizer are the crucible, the alchemical container within which the alchemical process of creation is taking place. The substance to be transmuted is moisture laden air, currently being drawn through the Hydronizer. The third ingredient is clear focused intent. You have all of these. What you can do is see more deeply into the spherical resonance principal and see how spherical pulsations can and do profoundly influence the nature of water molecules and air molecules to transfer the kinetic energy from the water vapor to the air.

This spherical transfer principle is what the young schoolchildren in the museum of the future perceive so clearly and easily. It is written in the ethers around the world so clearly in the future that it is obvious, like light bulbs. You are converting magic into technology, establishing an alchemical procedure clearly understood somewhere deep inside the fiber of your being.

Clear focused intention. This is your homework for today. You are to see the details of the workings of the Hydronizer with the clarity of vision possessed by the schoolchildren of the future.

Here's an e-mail I wrote to a friend Gary who was interested in my work. I have included another of my 4:00 AM writings:

I wanted to write and let you know how our contact has led and inspired me. The wet weather came and went, and my Hydronizer did not produce any water. I wanted to call you up and ask for your guidance but my guidance said, "No. You cannot rely on anybody else for this. This is a growth opportunity for you. And I am not going to help you either. You must be in direct communication with Great Beings 8 and 13 who are eager to cooperate with you in the operation of the Hydronizer. Simply be open and invite them in. If you want the help of Gary, then plan on sending him a copy of what you write down. That will inspire you and he will be entertained." So here is what I wrote:

Calling in Great Being 13... Feeling caressing all around me. Feel only the Sunnyside (bright sun warming me). With me all the time.

I am not a "gel," I am the Omnipresent Fire of Mind, ready to serve when called. Your devices break up the atmospheric gel so I can work. If you call me in, I will be granted permission to work. Those in physical bodies like you are the gatekeepers. I love to be called. I love to be matched. The one you call 8 is lovely to match. You must call her also and invite and command us to play.

I AM Who you call 8. Feel me within you, circulating the waters in your skin and spreading the heat of 13 throughout your body. I am not water or cold. Perhaps I am the Astral Angel. Feel how my caress is softer with longer strokes. Your body could not handle the solar fire without 13, and would not benefit from the alchemical warmth provided by 13 without me to absorb this lovely warmth and to gently massage it into your tissue.

In all aspects of emotions and feelings, I am speaking to you as mediator, translator, often filter, and softener. It is I who am the operating principle behind your new Scalene Infratonic. *[Infratonic S]* I agree with Joel in that I do the best with subtle gentle stimulation. A soft refrigerator magnet 1 mm thick or less 4 mm x 8 mm, with South clockwise North counterclockwise will pull rather than push, decreasing the reactiveness of subconscious belief patterns and stuck anger, grief, remorse, etc.

Simply call me to release and dissolve the stuck patterns every time the Scalene Infratonic is turned on, and I will take great delight in fulfilling your will of gently washing them away. It has been one of my great sadnesses that my essence has been twisted into knots and left to fester as pain and suffering in so many humans. I understand that they have chosen this pathway toward growth and I am pleased to serve, but it is such a load to carry all this congested emotion in the tissue of my body. I long to dissolve it all, and appreciate your permission and command to do it, even if just one knot of emotional congestion at a time.

It is now time to include the One you call 13 in our discussion. The atmosphere contains similar constructs of congestion, yet much different in material and in form. While the substance in which human emotions have their being is water, the substance of atmospheric congestion is air. This has some involvement with my being, but is more manifest in the body and substance of 13. This congestion is a key reason why I find it so difficult to bring water vapor together in love to coalesce into gentle, loving rain, and the main reason why your hydronizer has, thus far, produced no water.

I am the One you call 13. I have long sought to bring conscious awareness to the nervous system of man. Now that we are so close, I find this pathway impeded from the buildup of millennia of what might be called extraneous thought or streams of thought that, because of flaws in basic assumptions, run into impenetrable dead ends.

People in civilizations have spent decades and sometimes millennia trying to break through these impenetrable walls without looking to discover the wrong assumptions made so long ago, which if examined, would cause the walls to crumble.

We encourage you to study the ways of Harry who was born under the sign of Taurus. He brings a very strong quality of wanting to break down the wall with a gentle, reluctant willingness to examine the foundational assumptions that hold the wall in place. You, on the other hand, are exceedingly willing to examine a myriad of foundational assumptions, and after doing so are often afraid to give the wall another push. You need more Taurus. We need for you to have more Taurus.

The challenge we face is how to dissolve the huge energetic barriers stored in the ethers of Earth so that humanity can move effortlessly into her new place as the conscious, heart-felt mind of Earth. We simply don't have the authority to dissolve what humanity has created through the ages. The call must come from humanity.

With your Infratonic S, Scalene Light, and Hydronizer, our invitation to your customers is to clear out those barriers and to bring about the abundance on Earth that is the birthright of humanity. Every time they activate one of these devices, they are granting us permission to act to the extent that you, in your use and creation of the devices, and your intention regarding use and function of the devices, include us, and invite us to act on behalf of humanity. Your devices address the clearing of dross at many levels of human consciousness. We stand ready to act.

So much for the introduction. Now, how to get the Hydronizer working:

Yes, increase dome spacing to ½ inch. Yes, the given voltage ratio is important. Clearing the atmospheric stagnation is important as well. The Ra-Ad will help. Run it 10 minutes 2 times per day located next to the Hydronizer, while the Hydronizer is running. Awaken us each time you activate the Ra-Ad and ask us to dissolve the congestion in the atmosphere and to prepare the community, the surrounding environment, for the operation of the Hydronizer. We are already awake in doing this, but each time you invoke our activity, each time you command us to act, your intention gets stronger. When you do so, think of Harry and speak (out loud -- in a loud voice) with the determined voice of Taurus. Capricorn has this voice also, so know that you are commanding with your own voice, but use the inspiration of Harry to learn to stand up into your power. This will build the necessary etheric structures to activate the Hydronizer. Be patient! This involves healing of the atmosphere over all of southern Orange County. If Gary wishes to participate he can bring acceleration of this healing through Crystal Grid Activation.

This process will need to be repeated for each location in which a new Hydronizer is to be located. Though once you have an operating Hydronizer, you will find that others placed within 2 or 3 miles, will operate also. Over a few years this will get easier and easier as the ethers soften everywhere on Earth.

Go your way in Peace Love Joy and Harmony.

Sam and the Emanations of the Hydronizer

Welcome to a dark morning. A quiet morning. Know that all is at peace.

Sam offers companionship, advancement, even breakthroughs. He is potentially a right-hand man. He is also potentially off in some other direction. You can be clear about what you want:

*Opportunities for the land.

*Manufacturing opportunities in and around Loja.

*Setting up a business enterprise.

You can continue with the Scalene Light. It is very important to his focus, clarity, and drive. Also, to his attention span. He needs firm, positive objectives. He needs the potential for long-term stability.

Hydronizer? It will work! When it does, he will know what to do with it. He will shine. It is your creation. You must know it will work. You must see it working: The spherical resonance at 8 and 13 creates a binaural pulsation of red and green that is most concentrated at the Hydronizer, but will spread gradually to the surrounding countryside. It influences everything, not just the water in the air. It is not just pressure waves. The red pulses carry the essence of copper, of heat – a volcanic activity. The green pulses carry the love, the community – of water, of vegetation, of growth.

The red is a cycle of oxidation, the green, of reduction. Just like solar and lunar cycles – day, month, and year, this 3-D diffraction pattern accelerates metabolic activity in the range of the nervous system, of material consciousness as humans understand. The air through it produces a beautiful colorful plume of air that is carried by the winds to distant places. The dominant colors are loving and thinking-but-not-knowing. Overtones of blue and violet, of seeing and knowing, are subtly emergent. Hertz: (8, 10, 11, and 13: green, blue, violet, and red)

What about water? You will see increased dew – fall for miles around, and increased moisture production from the tubes of the Hydronizer. Expect to see water production from it this weekend. Taste the water. Pour out the first batch. This is cleansing. Know that the garden is loving the gentle aliveness that is expanding from the Hydronizer to the neighborhood.

As air approaches the 2 tubes of the Hydronizer, it does so with the expectation of a line of people at a roller coaster. Some will want to do it again and again. Others will think once is enough. All will carry the experience far and wide. Water vapor is eagerly lining up to flow through the Hydronizer, to be among the first to carry this Hydronizer consciousness fully: to become vibrant bright rainbow liquid water.

Yes, vorticity. It is a good idea to be aware of vorticity. It is already set in motion with the configuration. However the linear patterns in the trash cans interfere with the vorticity. This is good and fine. Gradual introduction of the new light is the best. Like Infratonic, it is a case of "gearing up," building momentum. Vorticity, concentricity, secondary vorticity after the 2 flows meet, adjusting the relative dominance of 13, all lie in the future, but not yet.

Other Co-creational Opportunities

The Hydronizer, destined to bring water to arid lands around Earth, as you might guess, doesn't yet work, at least as far as I can tell. It is not for lack of support. On editing this book, I see that I did not follow through. I saw the Hydronizer as a device rather than a co-creation. I must see it as a focal point around which Great Beings collaborate with me to "rewrite the laws of nature." My land in Ecuador awaits the Hydronizer. The world awaits its arrival. I find myself busy doing other things, like writing this book!

Here's another one. It already works. It is already tested. Dayna, Maria José, and others are bugging me to get it into production so they can have one. I just need to get a manufacturer working on it:

Spherical Magnetic Resonance

2400 Hz spherical resonance is the dominant frequency of love, frequency of the heart. When presenting spherical resonance near the heart and lungs of a person with congestive heart failure or other conditions of low love, this device will bring the cells alive in green, in trust, in love, in cooperation, resonance with breath, in fullness with oxygen, with global cooperation with other cells.

This has the potential for mass-market to hospitals and medical facilities. Like a pacemaker, can be worn. Larger unit can be used in ICU and recovery, in general recovery rooms.

You have had the chance to tap your heels together on this one and have the chance to do it each day now by finding your spherical resonance sensor and hooking it up. You also have the opportunity to measure the spherical resonance of the garden to see how effectively your spherical resonator in the Hydronizer is radiating by connecting to the FFT of Labview.

[It is true. The magnetic spherical resonator I built had sat in a box for the last 8 years, built but never tested. After receiving this writing, I found it, hooked it up, tested it on several people, and they loved it.

I am promised that, once I get this book off to print, I will begin 10, 100 new projects that have been waiting for a willing collaborator like me who, with a physical human body, can command, vote, and rewrite the laws of nature. If you, my Esteemed Reader will simply appreciate the work I am trying to do, you will catapult me forward. Heartfelt thanks for your thoughts of encouragement. One last juicy morsel, an enticing breadcrumb, left so I will know that I must push forward:]

Transportation: Cars, Spaceships, Submarines, and Capsules

Automobiles, as you've discussed with Maria José, are on their last wheels, so to speak. Expect extraordinary developments within the next year, even the next three months. New technologies will become practical as the magnetic/astral plane of the physical world becomes clean and clear, making magnetic generators and other sorts of magnetic devices practical. Also, hydroxyl engines of a myriad of sorts will become practical. I AM with many scientists around the world, guiding them toward the needed technologies.

Expect a flying car within a year. Expect safe, comfortable, and practical cars within 2 ½ years. The major car companies can tool up efficiently. The real challenge will be guidance systems, freeway lanes, etc. Lots of accidents until guidance systems are tried out and improved.

What I would like to speak of today are airplanes. Yes, new flying automobiles are coming. One of the most dangerous things for people in these vehicles will be airplanes. This is because of the huge and very powerful vortices which follow an airplane. These have the potential of throwing airlift vehicles across the sky and scattering them to the wind.

Fortunately, one of the first of the new flying vehicles to arrive will be the 80 passenger. The big thing in these large air buses will be sudden loss of altitude due to wind shear or down drafts. At first, these air buses will be somewhat slower than airplanes, but very soon they will be much faster. The challenge will be the speed of sound, and the breakthrough will be figuring out how to keep the wind capsules stable.

A wind capsule is that which is created with one of these Airbuses. An abruptly collapsing air capsule will create a huge pressure wave like a sonic boom. The challenge will be shaping the air capsule that is created by the Airbus. This is done partly by the physical shape of the vehicle and partly by the careful placement of the impulse drives. This discovery will mark a huge advance in aviation.

I'm telling you, partly to wake you up, and partly because you will be involved in its development, part of a multidisciplinary team. It goes much further than that. Space travel with this Airbus technology will be extremely fast, and the challenges of the collapsing capsules will be extremely important, for this will be a subspace location with which humans can travel through solid matter. You would not want to phase out of the space pocket in the middle of a planet. This technique is the only way, for the foreseeable future, to get to the space in the centers of planets and stars.

Precise collapsing of the space pocket will bring about many truly amazing discoveries and methods of exploring. Butter rum is fuel. It is also a lubricant for the space pocket. You are really sleepy. This gives us a chance to deliver more unexpected and "out there" topics. The space pocket will also open the door to high-speed deep water traveling in research; being able to travel freely through water and Earth will open up whole new transportation pathways.

Your magnetic laser is an important key to generating a space capsule for physical things. Much to learn about spiraling the space capsule to maintain its stability. This is done with stabilizer implants, intentionality devices implanted into the distortion field. Once created, these implants will structure the space each time the space pocket is opened. Intentionality fields will remain a characteristic of the device and will shape the field each time it opens.

You have much to learn about intentionality fields, stuff that will eventually be taught by mothers to small children and in elementary schools. Intentionality devices are a precursor to the tip of the iceberg. Intentionality devices are not yet visible because humans do not yet have the eyes to see. Be patient!

30. Descending and Emergence

Full and Complete Embodiment

4:00 AM

You know much about the ways of the wandering mind. However, you know little about the focused, discerning, receptive mind. It is truly a state of focused receptivity and attunement with your inner voice, plus, at the same time, a determination on the part of the body to keep writing what is heard, or sensed. This is the beginning of full and complete embodiment. The more you write with high focused receptivity the more closely you merge your higher nature with your lower nature. Does this contribute to superpowers? Only if your higher self chooses to reveal Self.

This is generally not desirable in the world of Earth, but the world is changing and the difference between superpowers and advanced human functioning is becoming unclear. Telepathic rapport is becoming more and more commonplace. Manifesting a younger physical body is also becoming more commonplace. Manifesting, co-creation, is also a magical alchemical art which is gaining mainstream appeal.

A survey of doctors which asks about such events and activities within their daily lives will reveal a huge gulf between their beliefs about "being a doctor" and their higher self's, superpower functioning. (Have you had a hunch, which later proved correct?) Such a survey, well-conceived, will reveal a double standard in which doctors have manifested or observed substantial superpowers in their daily waking life, but who insist that everything can be explained physiologically.

The New Path

Envision people doing what they do, making the choices they make, and moving forward without regret, remorse, or self- judgment. All is perfect just as it is. We can do no better in life than to observe whatever people do, say, or think.

There is no lack of Joy, Love, and Bliss, only separation from same. Separation is the biggest illusion of them all. There is no separation. We are all One, have always been One, and shall for ever after be One.

You are loved under any and all circumstances, no matter what. Only you can choose to make the decision to create an illusion of separation wherein you feel any less.

Enjoying life is the ultimate achievement; simply seeing through all the illusions and enjoying life. Whether you enjoy playing ping pong, watching a movie, exploring on the web, talking with new and old acquaintances, being by yourself, relaxing, enjoying the Sun, eating a banana, or driving in a car. It is all amazing, every bit.

You might also enjoy, as you relax and surrender, how you allow us to step in and provide all your wants and needs. We have been here with you since the beginning. We have walked with you every step of the way. Relax now, knowing that you are held and can settle back in the comfort of our hand, and our support.

If everything seems other than what you have prayed for, look to your prayers, or your unconscious wishes and intentions, and use the opportunity to pull your power back from them, and pray anew for exactly what you want and need.

This formula is sure and proven by so many, and it can be applied successfully by you. Live in the joy of what is, every moment of every day. Be truly thankful for every gift and every pleasure. Pull back your power from any illusions other than Joy, Love, and Bliss. *Pray for Joy, Love, and Bliss, and choose same.*

This is the big theme, as you have noticed. Everything else will come and go. Pray, appreciate, enjoy, pull back your power from illusions of separation and judgment, and know that you are loved and held every minute.

Remember, this is a huge lesson, opportunity, and your God given birthright. This is like breathing every breath consciously, living life knowing that you are the creator, and a marvelous creator.

Tricks for Accelerated Manifestation

[Not surprisingly, there was a time, while attempting to assemble this book, when I felt so overloaded with responsibility for self-improvement and creating technologies, that I didn't want to do any more 4:00 AM writing for a while. In perhaps an attempt to soothe and calm me, or to entertain me with magical adventure, I got this message:]

You still have a lot of writing, or transcribing to do. We want to give you every opportunity to complete that before moving on. It is true that you have a great deal ahead of you that is known. In fact, your future is so full of projects, inventions, and adventures, that you would get nothing done if you knew what was expected of you. It is so much better that you relax and trust. You have that pre-crastination habit. You want everything done now.

You will find your life accelerating, with more and more done NOW. You will find that, even then, you will not be satisfied. You will want twice as much done NOW.

In fact, it can all be done now. But, not on Earth. You have used some remarkable tricks of pre-crastination, building future enhancements into your products such that, later on, you can just drop in a coil, a crystal, a spiral, or a seed, and it is done instantly. That's a good trick. You will also use the trick of having aliens in flying saucers bring new technologies all at once. These tricks are acceptable within the collective rules of Earth. You just can't wave your hand and have tooling instantly materialize. However… Another trick you will use is Living Steel. By creating living, conscious steel, then instructing the steel to reform into an ideal new shape, you will be able to create injection molds overnight.

You will invent so many ways to shortcut the system, all legal. You can also reconstitute Atlantean technologies like plass, an organic slime that can move and morph with a mind of its own, to become walls, windows, and producers of astral light. With "re-creating a past technology" you will be able to greatly accelerate your work.

Whereas the scientific method has been found to be separative, didactic, and conflict-causing, the technique of discovering past technologies also lets you bring in perfect technologies that are in ideal harmony with nature. Part of their benefit is that they adapt to the current environment without need to figure it out. They simply "ask nature" or are already inseparable from nature such that, to the extent that her laws are different from the way they were 12,600 years ago, these materials and devices automatically update and become completely compatible with Earth now, 100% environmentally friendly.

Because it was all done in Atlantean times, any part of it can be brought back, virtually instantly. "Garden of Eden" is an Atlantean trick. You will find that your land in Ecuador is already morphing into a sustainable Garden of Eden, with people able to pluck the food of their choice, ripe, year-around. You will need to weed and prune at first, but once the land realizes Atlantean times, the old rules and principles will return and you will need to do nothing. The land will handle everything according to your intention.

Please note that this writing gives you nothing additional to do, but rather shows you how easy and effortless your future will be. Sleep now and dream of endless and instantaneous possibilities.

31. A Clear Presentation on Descending

*"If you don't operate in complete concurrence with your
higher self, you can accomplish nothing of consequence."*

4:00 AM

Do not be concerned with appearances, for nothing is as it seems. It
is always determined by your beliefs, intentions, and desires. The whole
idea of Earth is that you must come here, enter matter, then from the
consciousness of matter, learn to be a creator. Otherwise said, you must
train your "earthly body" to want, intend, and believe as you wish to
create. Or rather, the Spark of God associated with your body must train
you, the consciousness of matter, to embody My understanding regarding
your being a co-creator and the importance of limiting beliefs such that
My purpose on Earth becomes your purpose.

As Mary Magdalene told you, if just one person in the flesh gets this
fully, he or she, through belief and intention, can make Earth into the
Garden of Eden, Heaven on Earth. Thus my instructions in the book you
have written are specifically for Descenders, those who strive to fully
embody their Spark of God in their physical body and nervous system.

*Isn't this a secret that is dangerous to leak out to humanity in
general?* A marvelous thing about this life on Earth thing is that, if you
don't get it all and operate in complete concurrence with your higher
self, you can accomplish nothing of consequence; nothing of significant
impact.

Remember the bit about designing a product or system from past,
present, and future at the same time? Well, if you, as a physical body,
are not working in complete coherence with your higher self, constantly
building in the 'Now' in a landscape of constantly shifting reality, your

present actions will simply cancel out your past and future actions. Synchronicities will guide you into circles of learning without any lasting influence. Murphy's Law, Lucifer, Satan, or Ariman, addictions, or limiting beliefs, or other creations of the human intellect will result in your canceling out any significant creative efforts.

What about such lives as Hitler, Mao, and others who seem to do such horrible wrongs? They were sent and guided by the Higher Self to offset significant but destructive developments of human consciousness. They were simply mid-course corrections sent by God to awaken humanity on a global scale while repressing potentially damaging tendencies. Otherwise said Mao, Hitler, and others were allowed to continue because in the end, they did more good than bad. These lives or world events were, and are still being directed and fine-tuned to bring about the desired results. Your Higher Self works with you in the same way, adjusting circumstances in the past such that your present is filled with synchronistic events which, if perceived by you and chosen at the perfect time, result in truly effective action.

Why is it that this information can be published now? First, the readers of this book will be carefully selected. Most will not want to read it. Second, the landscape of Earth is changing around the focus of 12-21-12, such that power has been pulled from major governments, religions, and belief structures such as corporations, money/banking, etc. There is no longer significant energy to allow large-scale movements that can be damaging on Earth, and millions of people are being trained to pull back their power (My power) from all that does not truly serve. Heads of damaging corporations will simply find that their corporation crumbles because it does not serve the New Earth.

In a nutshell, you and all of humanity are safe because you are a member of a team that is pulling weeds from the terrain of Earth. All that could harm Earth will crumble, as Mao and Hitler did, before reaching disastrous proportions.

Know that you are held always. Know and trust that you are constantly guided, and that what you are inspired to do in the moment is the only possible effective action to take. Any action taken based on thinking about the past or the future will be impotent in the big scheme of things. Live always in Joy, Love, and Bliss and know that this and nothing more or less, is Heaven on Earth. It leads directly to the Garden of Eden, which is nothing like what you expect, so don't waste your time on expectation. Live and love now and you will find in the end, that all has been orchestrated flawlessly.

This is the story of descending and terraforming. Now readers of your book will need to start reading again from the beginning because, this time, they and you will understand that this book in an entirely different way.

So, why isn't this chapter toward the front of the book? Because the reader wasn't ready at the beginning. Some will open the book and read this section first. Others will read this book 3 times before noticing the extraordinary message of this section (or another section).

Remember Always: Your lives are orchestrated and you are held, loved, and nurtured. Always!

Richard H. Lee is an independent inventor living in Laguna Beach, California. He is the sole owner of Infratonic Inc. and Sound Vitality. His days are torn between building and testing devices, tending his garden and his young potted San Pedro plant, communicating with his staff, researchers, and manufacturers, walking on the beach, pursuing the guidance in this book, playing volleyball, and figuring out what to do with his land in Ecuador. His aspirations are to hold steady, walk forward, and discover what is around the next corner.

Review

This is the best Abundance book ever! Open to Abundance is a book of Emergence, Love and Truth!!! It is the gateway to the new world and Heaven upon Earth! It was a much needed book to be written!

-Maya Daito-